Brain, Nature, Mind, Universe: Interesting Facts & Unsolved Mysteries of Science and the Cosmos

Updated for 2024 with the Latest Scientific Discoveries

Oliver Elliott

Contents

Introduction 1

1. Record-breaking Animals 3

1.1 The Fastest Creatures

1.2 Incredible Vision

1.3 Keenest Sense of Smell

1.4 Best Hearing

1.5 Keenest Sense of Touch

1.6 Best Taste

1.7 Longest Lifespans

1.8 Giants and Miniatures

1.9 Amazing Adaptations

2. Extraordinary Plants 19

2.1 The Largest Plants

2.2 The Smallest Plants

2.3 The Oldest Plants

2.4 The Most Toxic Plants

2.5 The Most Resilient Plants

2.6 The Fastest-Growing Plants

2.7 The Most Colorful Plants

2.8 The Most Unusual Adaptations

2.9 Amazing Crops

3. Extreme Natural Phenomena 29

3.1 The Fastest Winds

3.2 The Windiest Places

3.3 The Heaviest Rainfall
3.4 The Heaviest Snowfalls
3.5 The Driest Locations
3.6 The Largest Deserts
3.7 The Coldest Temperatures
3.8 The Hottest Temperatures
3.9 The Extreme Earthquakes
3.10 The Tallest Waterfalls
3.11 The Longest Rivers
3.12 The Deepest Oceans
3.13 The Extreme Lakes
3.14 The Largest Volcanoes
3.15 The Longest Valleys
3.16 The Extraordinary Flatlands
3.17 The Amazing Basins
3.18 The Vital Rainforests
3.19 The Longest Lightning Strikes
3.20 The Most Frequent Thunderstorms
3.21 The Highest Tides
3.22 The Largest Waves
3.23 The Most Dangerous Places
3.24 The Longest Natural Hiking Trails

4. Unexplained Natural Events and Mysteries 47
4.1 The Taos Hum: A Sound That Drives People Crazy
4.2 The Bermuda Triangle: Mysterious Disappearances of Ships and Planes
4.3 Ball Lightning: A Mysterious Floating Light
4.4 The Tunguska Event: The Largest Impact Recorded in History
4.5 Hessdalen Lights: The Most Mysterious Lights on Earth
4.6 Lake Anjikuni Disappearance: The Unsolved Vanishing of an Inuit Village
4.7 The Mary Celeste: The Ghost Ship
4.8 Earthquake Lights: Mysterious Lights Accompanying The Shake
4.9 The Unexplained Forests

5. The Fascinating Universe 56
5.1 The Wonders of Gravity

5.2 The Fascinating Nature of Time: Is It An Illusion?
5.3 The Mysterious Black Holes
5.4 Dark Matter And Dark Energy: The Universe Is Mostly Dark To Us
5.5 The Fascinating Night Sky: A Closer Look
5.6 Heavenly Visitors: Spectacular Comets and Meteor Showers
5.7 Between Light and Shadow: Fun Facts About Solar and Lunar Eclipses
5.8 Unwelcome Visitors: Asteroid Impacts That Have Caused Disasters
5.9 Galaxies With Radio Signals: Are We Alone In the Universe?
5.10 Interstellar Migration: The Search for Habitable Planets
5.11 Everything Is in Motion: We Move Fast Through Space
5.12 Supernova: Fireworks in the Universe
5.13 The Origin and Fate of the Universe: Where Will It All End?

6. The Mysterious Brain 77
6.1 Brains: The Most Powerful Computers We Ever Know
6.2 Male and Female Brains: Do Brains Have a Sex?
6.3 Extraordinary Brain Capacities: More Powerful Than We Realize
6.4 Mysterious Brain Abnormalities
6.5 Brain Enhancement with Artificial Intelligence (AI): Friend or Foe?

7. The Elusive Yet Powerful Mind 111
7.1 The Unbelievable Mind-Body Connection
7.2 The Unconscious Mind: Operating Beyond Awareness
7.3 The Quantum Puzzle of the Mind
7.4 Mind and the Universe: Not Separate but Interconnected
7.5 Mind Over Matter: The Science on the Power of Thought
7.6 Consciousness After Death: Science or Fantasy?

8. Concluding Remarks 135

Stay Connected 140

Introduction

This is a book I've long dreamed of writing—not just to share the extraordinary wonders of nature and science, but because the process brings back the joy of discovery I felt in graduate school, immersed in the world of scientific literature. At our core, we are born curious about the universe, and this curiosity is one of our greatest gifts. Yet, as life pulls us into the routine and the mundane, this natural curiosity often fades. The wonders of nature and science have the power to reignite it, to remind us of the fascinating universe we live in.

The book is divided into eight chapters. The first three chapters reveal animals, plants, and natural phenomena of extraordinary qualities. Imagine a bird that dives faster than an aircraft at takeoff, a plant sprawling across an area larger than Washington, D.C., or a volcano as massive as New Mexico—each one a testament to nature's astonishing creativity.

Chapters 4 and 5 venture into the mysteries and uncharted territories of the natural world and the cosmos. From the Bermuda Triangle, where countless ships and planes have vanished without explanation, to black holes that warp space and time to their limits, these chapters remind us just how little we know in the face of the vast unknown.

Chapter 6 dives into the mystery of the human brain. This extraordinary organ is not only the most powerful computing device we know, capable of one quintillion (10^18) calculations per second, but it also shows the most puzzling symptoms when something goes wrong. Phantom limb sensations, blind patients who can unconsciously "see," and brain injuries that convince people their loved ones are imposters—all of these reveal just how much we have yet to understand about our brain.

Chapter 7 takes the exploration of the brain one step further—into the realm of the mind. Traditionally thought of as secondary to the physical brain, recent research is uncovering the mind's profound influence. From psychological tricks that promote weight loss, build strength, remove warts, and even reverse aging, meditation practices that reduce crime rates, to rigorous studies suggesting that human intention can affect physical reality, these findings challenge us to rethink the mind's role in shaping our world.

Finally, the last chapter concludes the book and reflects on my personal journey and motivation for writing this book.

This book is a journey into the wonders and mysteries of nature, science, and the cosmos. With a humble mind and an open heart, I invite you to join me on this adventure.

Chapter 1

Record-breaking Animals

1.1 The Fastest Creatures

Fastest Land Animals

- The African cheetah (*Acinonyx jubatus*) is the fastest land animal, reaching speeds of up to 75 mph (120 km/h). It can accelerate from 0 to 60 mph (96 km/h) in under three seconds but can only maintain these speeds for short distances (about 0.25 miles, or 400 meters) before needing to rest.

- The pronghorn (*Antilocapra americana*) is the second-fastest land animal, living in herds across North American grasslands and deserts. It can reach speeds of up to 55 mph (89 km/h) and sustain a speed of 40 mph (64 km/h) for at least 7 miles (11 km). It has large respiratory and cardiovascular systems relative to its body size to support its running ability.

- The springbok (*Antidorcas marsupialis*), the third-fastest land animal, is a small to medium-sized antelope found in the dry savannas and grasslands of southern Africa. It can run as fast as 55 mph (88 km/h) and is known for its "pronking" behavior, jumping over 10 feet (3 meters) into the air.

Fastest Birds (and of All Animals!)

- The peregrine falcon (*Falco peregrinus*) holds the record for the fastest animal in the world. It can see prey from a mile away and reach speeds of up to 240 mph (386 km/h, which is faster than an airplane at takeoff!) while diving to strike its prey. Peregrine falcons live and breed on every continent except Antarctica.

- Golden eagles (*Aquila chrysaetos*) can dive at speeds of up to 200 mph (320 km/h) to hunt small mammals, fish, birds, and reptiles. To achieve these speeds, they streamline their bodies by tucking in their wings and legs. Golden eagles are found in mountainous regions, open country, and rugged terrain across North America, Europe, Asia, and parts of North Africa.

- The white-throated needletail swift (*Hirundapus caudacutus*) is the fastest bird in flapping flight, reaching speeds of up to 105 mph (170 km/h). These swifts catch flying insects in midair with their wide beaks and migrate from Central Asia and southern Siberia to warmer regions like India, Southeast Asia, and Australia during winter.

Fastest Marine Animals

- The sailfish (*Istiophorus platypterus*) is the fastest marine animal and the fastest fish, swimming at speeds of up to 68 mph (109 km/h). Its sleek body and elongated bill minimize water resistance, while its sail-like dorsal fin can flatten to streamline its shape for rapid movement.

- The swordfish (*Xiphias gladius*) can reach speeds of up to 60 mph (97 km/h). It uses its sword-like bill to slice through schools of fish and stun its prey. As opportunistic hunters, swordfish rely on speed to catch various fish and squid. Contrary to popular belief, adult swordfish do not lose their scales; their scales are embedded deep in the skin, giving them a smooth, scaleless appearance.

- The blue marlin (*Makaira nigricans*) is the third-fastest fish, reaching speeds of 53 mph (85 km/h). With a powerful, muscular build, it easily outpaces prey such as smaller fish and squid. Blue marlins are not only fast but also large, growing up to 12 feet (3.6 meters) long and weighing up to 2,000 lbs (907 kg). Female marlins can live up to 27 years.

Fastest Insect

- Horseflies are the fastest flying insects. A male *Hybomitra hinei wrighti*, a species of horsefly in the family Tabanidae, was recorded reaching speeds of up to 90 mph (145 km/h) while chasing a female.

Humans v.s. Animals

- Usain Bolt set the world record of 9.58 seconds in the 100m race in 2009, reaching a peak speed of 27.8 mph (44.72 km/h). However, even this record-breaking speed places humans only among the top 30 fastest land mammals. Nevertheless, humans excel at endurance running compared to other animals, thanks to our long limbs, upright posture, and ability to sweat.

1.2 Incredible Vision

- Best Mammal Vision: Humans have exceptional vision among mammals, with the ability to see objects clearly at a distance. Human eyes have strong depth perception and can distinguish up to 1 million colors (isn't that amazing?). According to the snake detection theory, our advanced vision evolved partly due to the need to spot venomous snakes.

- Best Night Vision: Owls have remarkable night vision due to their high density of rod cells in their retinas (nearly a million per square millimeter), five times more than humans. Their eyes are particularly large relative to their body size, which enhances their ability to see in low light and makes them the best nocturnal hunters.

- Best Underwater Vision: Sharks have excellent underwater vision and can see up to 10 times better than humans underwater. They have a dense concentration of retinal rods and a reflective layer called the tapetum lucidum (Latin for "shining layer") in their eyes to amplify light in dark, murky waters.

- Broadest Field of Vision: Chameleons can rotate their eyes independently to have a nearly 360° field of vision—90° vertically and 180° horizontally. This allows them to observe their surroundings extensively without moving their heads, which is crucial for hunting prey without being detected.

- Best Color Vision: Bluebottle butterflies have at least 15 types of photoreceptors compared to the three in humans, making them the animals with the best color vision. Their photoreceptors can detect a spectrum of ultraviolet, violet, blue, blue-green, green, and red light. This allows them to detect specific colors in complex environments, such as hidden objects in vegetation.

- Most Complex Vision: The mantis shrimp has the most complex vision system in the animal kingdom. With 12 to 16 photoreceptors in their compound eyes, they can see ultraviolet and polarized light and can adjust their sensitivity to different wavelengths. This suggests that their eyes process visual information before it reaches the brain.

- Best Vision in the Animal Kingdom: Eagles have a visual acuity of 20/5, which is four times stronger than humans. They have near-panoramic vision and can spot prey from up to 3 miles away. They have enhanced color perception, and can also see UV light. They are the undisputed champions of eyesight in the animal kingdom!

1.3 Keenest Sense of Smell

- Bears are commonly thought to have the best sense of smell of any animal, with some estimates suggesting it's 2,100 times better than a human's. Their sense of smell is so good that it's difficult to measure, and estimates of their range vary widely. For example, some say a black bear can smell food from over a mile away, while others claim it can smell food from over two miles away. Polar bears' sense of smell is estimated to be even better, at up to 40 miles[1].

- African elephants likely have the most powerful sense of smell in terms of the sheer number of olfactory receptors. A research study showed they have an estimated 2,000 types of olfactory receptors, which allows them to detect a vast range of scents with incredible precision[2]. They can detect water sources from up to 20 km away, and they use the scent of other elephants to maintain social bonds.

- Bloodhounds are known for their exceptional sense of smell, which is estimated to be at least 1,000 times stronger than a human's. They have around 230 million olfactory cells in their large, sensitive scent membranes, which allows them to detect scents that humans can't. Bloodhounds can follow a scent trail for over 130 miles or a trail that's more than 12 days old. Bloodhounds' sense of smell has been used by law enforcement for nearly 200 years to track missing people and criminals. Their findings can even be used in court.

1.4 Best Hearing

- Pigeons have an extraordinary ability to hear infrasound, which are sounds with very low frequencies. While the human hearing range starts at around 20 Hz, pigeons can detect sounds as low as 0.05 Hz. This remarkable ability allows them to perceive infrasound emitted by distant storms, earthquakes, and volcanoes. Pigeons may use these low-frequency sounds as "sound maps" to navigate during long-distance flights. Their exceptional navigational skills were highly valued during wartime, and they were awarded 32 out of 53 PDSA Dickin Medals during World War II for their gallantry and devotion to duty[3].

- The greater wax moth has the most sensitive hearing in the insect world, despite its simple ear structure. With just a pair of eardrums and four receptor cells each, it can detect frequencies as high as 300,000 Hz[4] (compared with 20,000 Hz for humans), making it the animal with the best hearing known to science. This exceptional hearing allows the moth to effectively detect and evade predators like bats.

- Wolves possess remarkable hearing abilities, capable of detecting sounds from great distances—up to 6 miles (9.7 kilometers) away in forests and 10 miles (16.1 kilometers) in open areas. This exceptional hearing is vital for their hunting and survival in the wild.

1.5 Keenest Sense of Touch

- The star-nosed mole (*Condylura cristata*) is renowned for its extraordinary sense of touch. These moles have 22 movable, finger-like nasal appendages covered with more than 25,000 sensory receptors, giving them a sense of touch six times more sensitive than a human hand. They can even detect scents underwater by exhaling small air bubbles onto objects or scent trails and then inhaling the bubbles to carry the scent back through the nose.

- Tarantulas are large, hairy spiders with an exceptional sense of touch. Specialized hairs called trichobothria cover their legs and pedipalps, helping them detect faint air movements. These fine hairs can detect air movement as subtle as 10 millionths of a millimeter—about the size of two hydrogen atoms—allowing them to precisely determine the location of nearby creatures and whether they are potential prey or mates.

1.6 Best Taste

- Catfish have an extraordinary sense of taste. While an average person has about 10,000 taste buds, catfish have over 100,000 (some large catfish have up to 175,000), covering their entire bodies, including their whiskers, fins, and skin. These taste buds are highly sensitive and allow catfish to detect food from miles away, even in murky water. Catfish use their taste receptors to locate prey and can even determine the direction a taste is coming from. This exceptional sense of taste is crucial for catfish, which rely heavily on it for hunting and feeding. In contrast, chickens have the fewest taste buds of any animal, with only about 24.

1.7 Longest Lifespans

Longest-Lived Bird

- The oldest confirmed bird in captivity was a pink cockatoo named Cookie, who lived to be 83 years old at Chicago's Brookfield Zoo. Although some claim birds have lived over 100 years, these claims cannot be verified due to a lack of precise hatching records, especially for wild birds. In addition, bird enthusiasts sometimes exaggerate the ages of their birds. Cookie holds the Guinness World Record for the oldest confirmed bird.

Longest-lived Land Animal

- Jonathan, a 191-year-old Seychelles giant tortoise (as of August 2024), holds the Guinness World Record for the oldest living land animal. Born around 1832, Jonathan has lived through the terms of 40 US presidents and 31 governors of Saint Helena, his island home. Despite his advanced age and the loss of sight due to cataracts, Jonathan remains active and is hand-fed weekly to ensure he receives essential nutrients. His remarkable lifespan far exceeds the average 150 years for his species.

- Adwaita, meaning "one and only" in Sanskrit, was a male Aldabra giant tortoise residing at the Alipore Zoological Gardens in Kolkata, India. Weighing 250 kg (551 lb), Adwaita enjoyed a diet of wheat bran, carrots, lettuce, soaked chickpeas, and bread. In late 2005, a crack in his shell led to a fatal infection, resulting in liver failure and his death on March 22, 2006. At the time of his passing, Adwaita was believed to be at least 255 years old. If this estimate is accurate, Adwaita would hold the title of the longest-living land animal ever recorded.

Longest Lifespan in Mammals

- The bowhead whale (*Balaena mysticetus*) is considered the longest-living mammal, with a lifespan of over 200 years. These whales inhabit the Arctic and subarctic oceans and do not reach sexual maturity until around 25 years of age. Stone harpoon tips found in some harvested whales indicate that they can live well over 100 years, and some scientists believe they may live even longer.

Longest Lifespan in Vertebrates

- Greenland sharks are the longest-living vertebrates on Earth. Using carbon-14 dating, scientists estimated these sharks' lifespans to be around 400 years, with female Greenland sharks reaching sexual maturity around 156 years of age[5]. Their exceptionally long lifespans are likely due to the cold waters of the North Atlantic and Arctic Oceans, which slow down their metabolic and biochemical processes.

Longest-living Animals

- Glass sponges are considered among the longest-living creatures on Earth, with some individuals living for 10,000–15,000 years. A glass sponge in the Ross Sea is thought to be the oldest living animal, while another in the East China Sea is believed to have lived for 11,000 years. These sponges may have been alive during the last ice age. They are animals, not plants or rocks, and have complex, glass-like skeletons. They live attached to hard surfaces, filtering water to feed on bacteria and plankton.

- Immortal Jellyfish. Scientists discovered the *Turritopsis dohrnii*, known as the "immortal jellyfish," in the Mediterranean Sea in the 1880s. It has the unique ability to reset its life cycle, reverting to an earlier developmental stage when faced with physical damage or threats. Although tiny—just 4.5 mm across—the immortal jellyfish is also a resilient and increasingly aggressive species, spreading globally by hitching rides in ships' ballasts.

Longest-Lived Humans

- Although humans' lifespans are not impressive compared with some animals, we still have a decent lifespan among mammals. As of August 2024, the longest verified human lifespan is 122 years and 164 days, held by Jeanne Calment (1875–1997) of France. The longest verified lifespan for a man is 116 years and 54 days, held by Jiroemon Kimura (1897–2013) of Japan.

- Japan has the world's largest population of centenarians, with an estimated 146,000 people aged 100 and older. Japan's centenarian population has grown steadily over the past two decades, with a ratio of 7 female centenarians for every male centenarian.

1.8 Giants and Miniatures

Largest Mammal

- The blue whale (*Balaenoptera musculus ssp. intermedia*) is the largest mammal in the world and the largest animal and vertebrate that has ever lived. Antarctic blue whales can grow up to 98 feet (30 meters) long and weigh up to 440,000 pounds (200 tons). They have hearts the size of small cars, tongues the size of elephants, and arteries so wide that humans could swim through them. Blue whales are also the loudest animals on Earth, with calls that can reach 188 decibels. During feeding season, they can consume up to 7,936 pounds (3.6 tons) of krill per day.

- The African bush elephant (*Loxodonta africana*) is considered the largest land mammal in the world. Male African elephants can weigh up to 13,300 pounds (6 tons) and stand up to 13 feet (4 meters) tall, while females can weigh about 12,000 pounds (5.4 tons) and reach 14.7 feet (4.5 meters) tall. The largest elephant ever recorded was shot in Angola in 1974 and measured 35 feet (10.7 meters) from trunk to tail. African elephants are native to sub-Saharan Africa and live in herds in the savannas and forests. They are known for their long trunks and tusks, which can grow up to 8 feet (2.4 meters) long and are used for foraging.

Smallest Mammal

- The Etruscan shrew is the smallest living mammal by weight, weighing as little as 0.05 ounces (1.5 grams) and growing to be 1.42 to 2.05 inches (36 to 52 millimeters) long. It is found in Asia and the Mediterranean. Its heart beats up to 25 times per second, compared with 1.2 times per second in humans.

- The smallest mammal by body length is Kitti's hog-nosed bat, also known as the bumblebee bat. It is about 1.14 to 1.29 inches (29 to 33 millimeters) long and has the smallest skull of any mammal. It weighs around 0.05 to 0.07 ounces (1.7 to 2 grams), about the weight of two Skittles candies. It lives in caves in western Thailand and Myanmar in colonies of fewer than 100 bats.

Tallest Animal

- The giraffe (Giraffa camelopardalis) is the tallest land animal in the world, standing up to 18 feet (5.5 meters) tall. Their long necks, which can extend up to 6 feet, make up almost half of their height. Their legs are also taller than most adult humans. Giraffes are native to Africa and live mainly in savannas, where grasslands and trees provide the needed cover and food.

Largest Bird

- The common ostrich (*Struthio camelus*) is the world's largest living bird, reaching up to 9.2 feet (2.8 meters) tall and weighing over 346 pounds (156.8 kilograms). Native to Africa, ostriches inhabit grasslands, savannas, and shrublands. Although unable to fly, they use their broad wings (spanning up to 7 feet, or 2 meters) like sails to stabilize themselves while running. Their eggs are the largest of any bird species, averaging 3.1 pounds (1.4 kilograms) each and roughly equivalent to 24 chicken eggs.

Smallest Bird

- The bee hummingbird (Mellisuga helenae) is the world's smallest bird, measuring 2.25 inches (57 millimeters) long and weighing less than 2 grams (0.07 ounces). It is native to Cuba and is often mistaken for a bee. The female builds cup-shaped nests out of cobwebs and moss, and their eggs are about the size of coffee beans. The bee hummingbird's wings beat 80 times per second during flight and up to 200 times per second during courtship flights.

Largest Fish

- The whale shark (*Rhincodon typus*) is the largest fish in the world, growing up to 40 feet (12 meters) long and weighing up to 88,200 pounds (40 tons). Despite their large size and intimidating appearance, they are known for being docile and approachable.

Smallest Fish

- The dwarf minnow lives in the peat swamp forests of Southeast Asia and is considered the smallest fish in the world. Mature females are 0.31 to 0.41 inches (7.9 to 10.3 millimeters) long.

1.9 Amazing Adaptations

Mimicking Experts

- The mimic octopus, initially discovered in 1998 off the coast of Sulawesi, Indonesia, is the mimicking champion of the animal kingdom. Unlike other mimics that typically resemble only one species, this remarkable octopus can impersonate the appearance and behaviors of various creatures, including jellyfish, crabs, sea snakes, shrimps, and lionfish. The octopus mimics venomous lionfish and sea snakes to evade predators.

- The brown thrasher is likely the best mimic in North America in terms of sound because it can sing up to 2,000 different songs. It belongs to the Mimidae family and can effortlessly imitate the songs of other birds, such as the Chuck-will's-widow, wood thrush, and northern flicker.

- Like thrashers, northern mockingbirds also belong to the Mimidae family and they are versatile vocalists. They can produce a diverse range of sounds, from warbles and chirps to buzzes and a distinctive "chewp" note. They can learn and imitate up to 200 different sounds, including the calls of other birds like the eastern phoebe and seagulls, and even mimic human-made noises such as car alarms and creaky gates.

- European Starlings are also skilled mimics, imitating up to 20 different bird species, such as quail and killdeer. They can also mimic sounds made by motorcycles, tea kettles, car alarms, and phones, and even human speech. These birds are most abundant in Europe and are known for their adaptability and sometimes aggressive behavior. All starlings in North America descended from 100 birds released in New York's Central Park in the early 1890s.

Extreme Survivors

- Although many animals have adapted to their environments and developed excellent survival skills in extreme conditions, none compare to tardigrades. Tardigrades, also known as water bears, are tiny creatures (0.1–1.2 mm long) that inhabit various moist environments, such as moss, soil, and water. They can enter a state of suspended animation called cryptobiosis, in which they dehydrate to just 3% of their normal water content and reduce their metabolism to 0.01% of the normal level. In this state, known as the "tun state," they can withstand extreme conditions, including radiation 1,000 times stronger than what humans can endure, temperatures ranging from near absolute zero (–459.67°F or –273.15°C) to 300°F (149°C), the vacuum of space, and pressures six times greater than those found in the deepest ocean trenches (15,750 pounds per square inch, or 1,071 times the atmospheric pressure at sea level). They are the indestructible survivors of the animal kingdom.

- In comparison, while humans cannot survive under such extreme conditions, they can still show remarkable endurance. Wim Hof, known as "The Iceman," is a Dutch endurance athlete famous for his extraordinary ability to withstand extreme cold. He has achieved over 20 world records, including running a 26-mile marathon at -4°F (-20°C) barefoot and in shorts, and set a Guinness World Record by remaining submerged in ice for 1 hour, 52 minutes, and 42 seconds. In addition, he climbed Mount Kilimanjaro in just shorts and shoes within 48 hours, and he swam 188 feet (57.5 meters) under a frozen lake. Hof's accomplishments extend beyond cold resistance, and he completed a half marathon in the Namib Desert without drinking water. Hof's secret lies in the Wim Hof Method, which utilizes controlled breathing exercises, meditation, and mental focus to activate the body's stress response. This process releases natural painkillers like opioids and cannabinoids, inducing a euphoric state that reduces pain and cold sensations. MRI studies confirm that his brain effectively "hacks" physiological systems to suppress discomfort and enhance endurance[6].

Power Biters

- Saltwater crocodiles are the largest living reptiles, growing up to nearly 20 feet in length and living anywhere from 60 to 75 years. They have the strongest recorded bite force of any living animal at 3,689 pounds per square inch (psi), or 16,414 newtons[7]. This is equivalent to the weight of a small car. They can swim up to 18 miles per hour, three times faster than the fastest human swimmer.

- The American alligator (*Alligator mississippiensis*) is native to the southeastern United States and can exert a powerful bite force of 2,961 pounds of force, or 13,172 newtons[8]. It held the record for the most powerful bite force since 2006 until being surpassed by the saltwater crocodile in 2012. Despite this, its jaw muscles are relatively weak, and a human could hold its jaws closed with a hand or a strong rubber band.

- The shortfin mako shark (*Isurus oxyrinchus*) has the most powerful shark bite ever recorded, measured at 3,000 pounds of force, or 13,000 newtons. The measurement came from a shortfin mako near New Zealand in 2020. Although some studies suggest a great white shark and its extinct prehistoric relative, the megalodon, have an even more powerful bite, these claims are not verified by living recordings. The shortfin mako can grow to be 12 feet long and weigh over 1,200 pounds. It's also one of the fastest sharks and can reach speeds of up to 45 miles per hour.

- The hippopotamus (*Hippopotamus amphibious*) has a bite force of about 1,800 pounds of force (8,100 newtons), stronger than any other land animal. This is much stronger than a lion's bite force of 1,000 pounds of force (4,450 newtons), a polar bear's bite force of 1,120 pounds of force (5,000 newtons), and a jaguar's bite force of 1,500 pounds of force (6,500 newtons). Hippos can open their mouths 150° and have sharp triangular lower canines that can be 30–50 cm long. Hippos are herbivores, but they can be dangerous to humans if disturbed.

Nature's Nappers

- In the wild, big brown bats (*Eptesicus fuscus*) hibernate for 64–66 days, but one individual in captivity hibernated for 344 days. During hibernation, their heart rate drops from 1,000 to 4–62 beats per minute, and they may breathe only once every two hours. Their body temperature can decrease to around 5°C, and they experience greatly reduced metabolism and oxygen consumption. However, upon waking from hibernation, their heart rate can increase from about 12 to 800 beats per minute in a short time.

- Arctic ground squirrels spend seven to eight months in hibernation. They can enter a state of torpor and lower their body temperature from around 99°F (37°C) to as low as 27°F (-3°C), which is below freezing. This is the lowest body temperature ever recorded for a hibernating mammal. Such a low temperature allows their cells to survive when there's a shortage of blood and oxygen.

Always Awake

- Butterflies don't sleep like humans do, but they do rest in a behavior called "roosting." Butterflies may roost for a single night or for the entire winter season. Roosting in groups helps butterflies conserve energy and protects them from predators. Butterflies rest when it's cool or cloudy, or at night. Because butterflies don't have eyelids, they rest with their eyes open, usually hidden in leaves, bushes, or bark. They may hang upside down from leaves or twigs using their tarsi (feet). Their metabolic systems slow down significantly when they're resting.

- Dolphins' sleep differs from that of humans and other land mammals. They experience unihemispheric slow-wave sleep (USWS), where half of their brain remains awake. The active hemisphere monitors breathing, the environment, and signals when to surface for air, keeping one eye open during sleep. Dolphins switch active hemispheres every two hours, allowing each hemisphere approximately four hours of sleep within a 24-hour period. For protection from predators, dolphins often sleep in pods or with companions, taking turns sleeping or resting near each other.

- Common swifts hold the world record for continuous flight, staying airborne for up to 10 months a year, while alpine swifts manage up to six months non-stop. Swifts even sleep on the wing by alternating brief rest periods during flight. At night, they ascend to altitudes of up to 2,000 meters to avoid predators and obstacles, gliding at speeds of 12.4 miles per hour (20 km/h). Their unique intermittent wing-beating pattern—four seconds of activity followed by three seconds of rest—enables them to conserve energy and safely navigate without landing.

- Unlike animals, humans typically need at least seven hours of sleep to stay alert and function well during the day. However, there are exceptions. Daisuke Hori, a

40-year-old man from Tokyo, claims to have slept only 30 minutes a day for the past 15 years. This minimal sleep routine, he says, allows him to maintain a very active lifestyle, including working over 10 hours a day, exercising, and pursuing numerous hobbies. He attributes his ability to function on so little sleep to years of self-training and believes it has brought him exceptional happiness and productivity. Despite his claims, experts caution against such extreme sleep deprivation, as most people require 7-9 hours of sleep to maintain their health and well-being.

- Another striking case is that of Paul Kern, a Hungarian man who was injured during World War I in 1915. A bullet struck his head and lodged in his frontal lobe. After undergoing surgery, he awoke to find that he could no longer sleep. This condition lasted for 40 years, until his death in 1955. Despite living without sleep, he showed no major cognitive issues, a phenomenon that baffled scientists. Theories suggest the bullet may have damaged the sleep-regulating region of his brain, or that he may have experienced brief, unnoticed naps.

Venomous Legends

- The inland taipan snake is known for having the most toxic venom of any snake, and possibly any animal, in the world. The inland taipan's venom is a combination of enzymes and other factors that can paralyze nerve endings, destroy muscle tissue, and cause severe bleeding. The inland taipan's LD50 is 0.025 mg/kg, meaning that a single bite could kill nearly 250,000 mice or over 100 adult humans. Inland taipan snakes are found in remote areas of southwestern Queensland and northwestern South Australia.

- The Australian box jellyfish (*Chironex fleckeri*) is considered the most venomous marine animal in the world, and its sting can be lethal. Each jellyfish has about 60 tentacles up to 3 meters long, and each tentacle contains millions of venom-filled microscopic hooks. A single box jellyfish has enough venom to kill more than 60 people and can cause death within minutes. Its venom contains bioactive proteins that can cause various effects, including extreme pain, tissue destruction, paralysis, cardiac arrest, and death.

Storm Chasing Birds

- The Desertas petrel, a small seabird from the North Atlantic, has been found to chase cyclones. In a 2024 study, scientists used GPS to track 33 petrels and found that nearly a third of them actively flew toward cyclones when they were about 900 kilometers from the storm's eye. Some birds followed the cyclone's wake for days, traveling thousands of kilometers[9]. Scientists believe that the birds may use the low-frequency infrasound produced by wind and waves to detect cyclones from far away. The reason? A potential big food reward. Researchers discovered that cyclones churn cooler waters, leading to an increase in phytoplankton (free-floating, microscopic algae). Phytoplankton then attracts fish and cephalopods, which are the petrels' prey. The study suggests that petrels exploit the food-rich environment created by cyclones to feed themselves.

- Streaked shearwaters, seabirds that nest on islands off Japan, exhibit a surprising and unique behavior: flying into the eye of a typhoon to survive. Researchers tracking these birds with GPS over 11 years found that, instead of avoiding storms like other seabirds, some streaked shearwaters head toward the center of the cyclone and fly near the eye for hours. The study revealed that 13 out of 75 tracked shearwaters flew within 60 kilometers of the eye[10]. This area, also known as the "eye socket," has the strongest winds, sometimes reaching 75 kilometers per hour. These birds stayed in this zone for up to eight hours, tracking the storm's movement. This previously unseen behavior in any bird species helps shearwaters avoid being blown inland, where they could crash or be hit by debris. The birds were more likely to follow this strategy during stronger storms to avoid being pushed toward dangerous landmasses. Scientists believe such flying tactics could be more common among birds than previously thought.

Notes

1. *Bear Series, Part One: A Bear's Sense of Smell - Yosemite National Park (U.S. National Park Service).* (n.d.). https://www.nps.gov/yose/blogs/bear-series-part-one-a-bears-sense-of-smell.htm

2. Niimura, Y., Matsui, A., & Touhara, K. (2014). Extreme expansion of the olfactory receptor gene repertoire in African elephants and evolutionary dynamics of orthologous gene groups in 13 placental mammals. *Genome Research, 24*(9), 1485–1496. https://doi.org/10.1101/gr.169532.113

3. Chrisler, T. S., & Quine, D. B. (1980). *Can pigeons "Hear" their way home?* https://watchbird-ojs-tamu.tdl.org/watchbird/index.php/watchbird/article/view/1938

4. Fields, H. (2013). ScienceShot: Wax moth has most sensitive ears in insect world. *Science*. https://doi.org/10.1126/article.26109

5. Nielsen, J., Hedeholm, R. B., Heinemeier, J., Bushnell, P. G., Christiansen, J. S., Olsen, J., Ramsey, C. B., Brill, R. W., Simon, M., Steffensen, K. F., & Steffensen, J. F. (2016). Eye lens radiocarbon reveals centuries of longevity in the Greenland shark (Somniosus microcephalus). *Science*, *353*(6300), 702–704. https://doi.org/10.1126/science.aaf1703

6. Learn, J. R. (2018, May 21). Science explains how the Iceman resists extreme cold. *Smithsonian Magazine*. https://www.smithsonianmag.com/science-nature/science-explains-how-iceman-resists-extreme-cold-180969134/

7. Erickson, G. M., Gignac, P. M., Steppan, S. J., Lappin, A. K., Vliet, K. A., Brueggen, J. D., Inouye, B. D., Kledzik, D., & Webb, G. J. W. (2012). Insights into the Ecology and Evolutionary Success of Crocodilians Revealed through Bite-Force and Tooth-Pressure Experimentation. *PLoS ONE*, *7*(3), e31781. https://doi.org/10.1371/journal.pone.0031781

8. Erickson, G. M., Lappin, A. K., Parker, T., & Vliet, K. A. (2003). Comparison of bite-force performance between long-term captive and wild American alligators (*Alligator mississippiensis*). *Journal of Zoology*, *262*(1), 21–28. https://doi.org/10.1017/s0952836903004400

9. Polidoro, J. (2024, August 6). Why a small seabird dares to fly toward cyclones. *Science News*. https://www.sciencenews.org/article/seabird-fly-toward-cyclones-prey

10. Kreier, F. (2022, October 17). Some seabirds survive typhoons by flying into them. *Science News*. https://www.sciencenews.org/article/seabirds-typhoons-shearwaters-survival

Chapter 2

Extraordinary Plants

2.1 The Largest Plants

- Scientists discovered the world's largest plant in 2022, a seagrass called *Posidonia australis* in Shark Bay, Australia[1]. The plant stretches for over 112 miles (180 kilometers), an area larger than Washington, DC, and is estimated to be at least 4,500 years old. The plant's size is due to its ability to clone itself, creating genetically identical offshoots. This process, called polyploid clonality, allows organisms to outcompete their predecessors and occupy new environments.

- Pando (*Populus tremuloides*), a clonal colony of quaking aspen trees in Utah's Fishlake National Forest, is considered the largest living organism by mass. It weighs an estimated 13.2 million pounds (around 6,000 tons) and comprises 47,000 genetically identical stems spanning 106 acres (roughly 80 American football fields). Pando is three to five times heavier than the largest known redwoods and roughly as heavy as 35 blue whales or 1,000 elephants. All stems share an 80,000-year-old root system. Pando reproduces asexually through self-replication.

- The General Sherman Tree (*Sequoiadendron giganteum*) in Sequoia National Park's Giant Forest is the world's largest tree by volume. It stands 275 feet (84 meters) tall and has a 100-foot (30-meter) wide trunk. It's estimated to weigh around 2.7 million pounds (1,225 tons) and be approximately 2,100 years old.

- The *Rafflesia arnoldii*, also known as the corpse flower, is the world's largest individual flower. It can grow to be 3 feet across and weigh up to 15 pounds. It emits a foul odor that smells like rotting meat. The flower is native to the rainforests of Southeast Asia, including Indonesia, Malaysia, Thailand, and the Philippines.

A parasitic plant, it attaches itself to a host vine to obtain water and nutrients. The corpse flower is a rare, endangered plant that is almost never cultivated in botanical gardens.

2.2 The Smallest Plants

- Watermeal is the world's smallest flowering plant and measures a mere 0.004–0.008 inches (0.1–0.2 mm) in diameter. It belongs to the genus Wolffia, a type of duckweed. This bright green, oval-shaped plant grows in mats on the surface of calm freshwater bodies like ponds, lakes, and marshes. Watermeal is rich in protein and is an excellent food source for aquatic birds and fish. It is even used in Thai cuisine.

- The dwarf willow (*Salix herbacea*), also known as the snowbed willow or least willow, is often considered the world's smallest tree. It grows to a height of only 0.4–2.4 inches (1–6 cm). This tiny, ground-hugging willow is a member of the Salicaceae family and is well adapted to survive in the Arctic and subarctic tundra. The dwarf willow has both male and female catkins, but on separate plants. The female catkins are red, while the male catkins are yellow. However, some consider the dwarf willow to be a woody shrub rather than a tree, so the title of "smallest tree" is a matter of debate.

2.3 The Oldest Plants

- Estimated to be around 80,000 years old, the Pando quaking aspen colony is not only one of the largest plants on Earth, but also the oldest living plant discovered so far.

- King's holly (*Lomatia tasmanica*) is estimated to be at least 43,000 years old. It was discovered in 1937 by Tasmanian explorer Denny King, and is believed to be located somewhere in the wilderness of southern Tasmania. King's holly reproduces only vegetatively, meaning a new plant grows when a branch falls and develops its own roots.

- Methuselah, a Great Basin bristlecone pine tree located in the White Mountains of California, is estimated to be 4,856 years old as of August 2024[2]. It is the oldest known living non-clonal tree in the world. The tree's age was determined by counting the rings in core samples taken from its base in 1957. To protect it from vandalism, the U.S. Forest Service keeps the tree's exact location a secret. The tree is named after the biblical patriarch Methuselah, who was said to have lived 969 years.

2.4 The Most Toxic Plants

The Deadliest Bean

- Castor bean is a shrub native to Africa and is arguably the world's most toxic plant. Castor beans contain ricin, a toxin estimated to be 6,000 times more poisonous than cyanide and 12,000 times more poisonous than rattlesnake venom. One castor bean can be lethal to a child, while three or more can be lethal to an adult. Symptoms of ricin poisoning include vomiting, severe abdominal pain, diarrhea, and convulsions. Heat (176°F, or 80°C) inactivates the toxin.

- The jequirity bean is the seed of the rosary pea, a poisonous, evergreen vine native to India and other tropical and subtropical areas of the world. It gets its name from its rosary bead-like appearance. Jequirity beans are extremely toxic because they contain abrin, a toxin that's up to 75 times more potent than ricin, the poison found in castor beans. Abrin inhibits protein synthesis in cells and induces cell death. However, poisoning from jequirity beans is rare because the seeds have a hard shell that prevents the toxin from being digested in the gut.

The Deadliest Tree

- The manchineel tree (*Hippomane mancinella*) is considered the world's most toxic tree. Its apple-like fruits are the reason for its Spanish name, *manzanilla de la muerte*, meaning "little apple of death." All parts of the tree are toxic, including the sap, bark, and fruit. Contact with the tree's sap can cause severe blisters, and exposure to the sap can lead to blindness. Even standing under the tree during a rainstorm can cause blistering if raindrops contain sap. Eating the fruit can cause agony and potentially death. The manchineel tree is native to the Caribbean, Florida, and northern South America.

The Deadliest Mushroom (Which Is Not A Plant, Indeed)

- The death cap mushroom (*Amanita phalloides*) is the world's deadliest, responsible for 90% of mushroom-related fatalities. This medium to large mushroom has a greenish-gray cap, white gills, a white ring around its stem, and a large, white sac at the stem's base. Native to Eurasia, the death cap was accidentally introduced to North America in the 1930s. It's now found in urban environments and slowly colonizing the West Coast. The mushroom contains amatoxins, which cause life-threatening liver and kidney damage. A single mushroom can contain enough poison to kill an average adult. Symptoms of death cap poisoning include low blood pressure, nausea, and vomiting. Never touch this mushroom if you encounter it in the wild!

The Deadliest Plant in Europe and North America

- Wolf's bane, also known as monkshood, is one of the most poisonous plants in Europe. It contains a neurotoxin called aconitine, which can be absorbed through the skin and cause poisoning by contact. As little as 2 milligrams of aconitine or 1 gram of the plant can cause death from respiratory paralysis or heart failure. In ancient times, people used the plant on arrow tips and as bait to kill wolves (hence its name). It's native to western and central Europe, but there are about 100 species of wolfsbane, with five native to North America.

- Water hemlock is considered the most poisonous plant in North America. It contains a toxin called cicutoxin, which rapidly produces gastrointestinal symptoms such as nausea, vomiting, and abdominal pain within 60 minutes of ingestion. The toxin also acts directly on the central nervous system, causing seizures and other symptoms. The plant's root contains the highest concentration of the toxin, and even a single bite can be lethal. Water hemlock grows in wet, marshy places and is often confused with wild parsnips or celery.

The Deadliest Temptation

- Deadly nightshade contains compounds called atropine and solanine, which can be fatal if ingested. The berries of the deadly nightshade plant resemble blueberries, making them easy to ingest by accident. The plant's berries are particularly dangerous because they taste sweet but are deadly. Deadly nightshade has a long history of use and abuse, including as a poison and in cosmetics. The plant's name comes from the Italian word for "beautiful woman" because women used to apply its juice to their eyes to dilate their pupils. It is reported that as few as two berries can kill a child, and 10–15 can kill an adult.

2.5 The Most Resilient Plants

- The Welwitschia mirabilis, also known as the Welwitschia, is a unique plant that has survived for thousands of years in the harsh Namib Desert. Some individual Welwitschia plants are estimated to be between 1,500 and 2,000 years old. It's considered a "living fossil" and is believed to have evolved around the same time as dinosaurs. The Welwitschia's deep taproot, which can extend 5 feet (1.5 meters) or more, helps it survive in the arid desert by absorbing water from sea fog and deep groundwater.

- Resurrection plants (*Selaginella lepidophylla*), found in the Chihuahuan Desert of the United States and Mexico, are remarkably tough. They can lose nearly all their water and remain desiccated for years. However, they can quickly revive within 48 hours with a little water. Resurrection plants accumulate large amounts of sucrose and trehalose (two types of sugars) in their tissues, which stabilize enzymes and cellular structures in the absence of water and prevent damage from salts.

2.6 The Fastest-Growing Plants

- Bamboo is the fastest-growing plant in the world and can grow up to 2.91 feet (0.9 meters) per day, or 1.5 inches (38 millimeters) per hour. This is equivalent to adding about the size of a paperclip to the bamboo's height every hour! Bamboo only grows visibly above ground for about 60 days in the spring. After that, the canes stop growing, and the plant redirects its energy to its roots. Bamboo grows fast mainly because of its hollow stems and the rapid expansion of its cells after water absorption.

- Although not a plant, giant kelp (*Macrocystis pyrifera*) is an extremely large brown alga and one of the fastest-growing species in the ocean. It grows at an average rate of 11 inches (28 cm) per day but can grow up to 24 inches (61 cm) per day in ideal conditions. It can reach over 100 feet (30 meters) long in just over a year. Giant kelp thrives in turbulent water, which provides nutrients for its growth.

2.7 The Most Colorful Plants

The Most Colorful Tree

- The rainbow eucalyptus tree is known for its striking colors and patterns and is considered to be the most colorful tree in the world. The tree's bark is naturally brown but peels away in strips to reveal a bright green inner bark. As the green bark is exposed to air, it changes color over time, taking on shades of purple, blue, orange, red, and yellow. The tree sheds its bark at different times throughout the year, revealing new colors and patterns each time. Because of this, no two rainbow eucalyptus trees have the same pattern. The rainbow eucalyptus can grow up to 246 feet (75 meters) tall. It was originally brought to Hawaii in 1929 and can be found in various locations throughout the islands.

The Most Colorful Flowers

- Bird of paradise flowers (*Strelitzia reginae*) are known for their vibrant colors and striking appearance. They feature a range of colors, including orange, red, blue, and yellow. The flowers have intricate patterns and resemble a tropical sunset. Their shape is also notable, with petals that fan out like wings and a beak-like structure. Bird of paradise flowers are native to South Africa. They are popular houseplants and are less sensitive to cold than other tropical flowers.

- The blue lotus, also known as the sacred blue lily of the Nile, is a tropical water lily with star-shaped flowers that bloom during the day. The flowers can grow up to 4–6 inches (102–152 millimeters) across and have upright, pointed petals that spread flat on the water's surface. The blue lotus flower is known for its striking color, a vibrant sky blue with hints of purple. The pointed tips of the petals are sky blue, fading to white toward the center, which creates a glowing appearance. In ancient Egypt, the blue lotus was linked to rebirth and the sun god Ra.

2.8 The Most Unusual Adaptations

- Deadly catch. The Venus flytrap is a carnivorous plant with one of the fastest trapping mechanisms in the plant kingdom. It lures prey with a fruity scent and the reddish lining of its leaves, which have tiny hairs called trichomes that are sensitive to touch. When an insect touches the hairs, the plant's leaves snap shut in less than a second, trapping the prey inside. The plant then secretes enzymes that dissolve the prey's body, providing nutrients. The trap reopens after about 10 days to await its next meal.

- Stone-like plants. Lithops are small, succulent plants native to the deserts of southern Africa. They have evolved a pair of thick leaves that resemble colorful pebbles. The plant has a short stem, and its flowers emerge from a fissure at the top of the leaves. Its appearance effectively protects it from being eaten by animals, and lithops are very difficult to spot, even for a trained eye. Lithops can live up to 40 or 50 years.

- Tree of Life. The baobab tree is an African succulent that can store over 1,000 gallons of water in its thick trunk during the rainy season, giving it its iconic look. The tree's bark and fruit have over 300 life-sustaining applications. For example, its iron-rich leaves are edible. The fruit is a superfood with six times more vitamin C than oranges. The fruit juice can be used as a medicine to settle a child's stomach. The seeds can be used to create oils. The bark can be turned into rope and clothing. In addition, the tree is the root of many Indigenous remedies, traditions, and folklore. The baobab tree has been known to survive both drought and fire, and the oldest baobab is more than 2,500 years old.

- Painful stings. Stinging nettle is a plant frequently used in Western herbal medicine. Its leaves and stems have tiny, hollow hairs about 1 millimeter long. When the hairs touch the skin, their tips break off, injecting chemicals—such as histamine, acetylcholine, serotonin, and formic acid—into the skin. These chemicals cause an immediate painful skin reaction, which can include burning, itching, or tingling. The skin can also redden and develop a non-spreading rash. Strangely, stinging nettle can decrease pain in painful areas of the body. Scientists believe this is because it reduces inflammatory chemicals in the body and interferes with pain signal transmission.

- Touch me not. The sensitive plant, also known as *Mimosa pudica*, reacts to physical contact and darkness with rapid movements and changes in leaf orientation. When touched or shaken, the plant's leaves fold inward, and its branches droop—a reaction known as thigmonasty. This movement is thought to be a defense mechanism that startles herbivores. The leaves reopen after several minutes. The plant's quick response to touch is due to specialized cells at the bases of the leaves and leaf stalks that release water rapidly.

- The tree that walks. There is a rumor that the walking palm tree, also known as *Socratea exorrhiza*, can walk up to 66 feet (20 meters) a year[3]. The process includes three steps: soil erosion, which allows the tree to grow new roots toward the sun; root settlement, where the new roots settle in the new soil, and the tree bends toward them; and old root elimination, in which the older roots disintegrate, and the tree shifts its center of gravity toward the new root system. However, some scientists debate whether the walking palm tree actually walks. More careful studies will help ascertain the tree's walking ability in the Central and South American jungles.

2.9 Amazing Crops

- Deadly spicy. As of October 2023, the world's hottest pepper is Pepper X, which measures an average of 2.69 million Scoville Heat Units (SHU). This is about 1,000 times hotter than the average jalapeño pepper. The former record holder was the Carolina Reaper, with an average of 1.64 million SHU. Both peppers were developed by Ed Currie through cross-breeding. Eating a Pepper X can cause heartburn and severe abdominal cramps, and even touching it can cause searing pain to the skin. By comparison, a jalapeño pepper measures between 2,500 and 5,000 SHU, while a Thai pepper measures between 50,000 and 100,000 SHU.

- The sugar queen. The mango is considered the sweetest fruit in the world, with one mango (around 220 grams) containing a whopping 45 grams of sugar. The Carabao mango and Manila mango from the Philippines, and super mangoes from Egypt, are among the sweetest varieties. This sweetness comes from the mango's high fructose content. As runner-ups, a cup of grapes (150 grams) contains 23 grams of sugar, and a cup of cherries (140-150 grams) contains 18 grams. So watch out for the sugar content the next time you choose a snack!

- Fabulously Fat. Avocados are the fruit with the highest fat content, with about 80% of their calories coming from fat. A medium avocado typically contains around 22 to 23 grams of fat, roughly 250 calories. The fat in avocados is mostly monounsaturated, which helps lower LDL (bad) cholesterol levels and reduce the risk of heart disease and stroke. Avocados are also high in fiber and contain many vitamins and minerals.

- Titan of the Fields. The heaviest vegetable ever recorded (as of October 2023) is a 2,749-pound (1,246.9 kg) pumpkin grown by Travis Gienger in the USA[4]. It was presented at the 50th Safeway World Championship Pumpkin Weigh-Off in Half Moon Bay, California, on October 9, 2023. Gienger named the pumpkin "Michael Jordan" as a nod to the legendary basketball player. This pumpkin beat the previous record of 2,702 pounds (1,226 kg) set by Stefano Cutrupi in Italy in 2021. Gienger, a horticulture teacher, is known for growing giant pumpkins, including a 2,560-pound pumpkin in 2022 that was carved into an eagle-themed jack-o'-lantern.

- Can't live without. Rice is one of the world's most important crops, feeding more people than any other staple food. It is a principal food for over 3.5 billion people, particularly in Asia, Latin America, and parts of Africa—more than half of the global population. Rice cultivation began in ancient India or Southeast Asia, and today, China, India, and Indonesia lead the world in rice production. This versatile grain provides over 20% of all calories consumed by humans.

- The oldest crop. Taro is considered the oldest cultivated crop in the world. According to research published in 2023, its use dates back nearly 28,000 years in the Solomon Islands, far older than the lentil's 13,000 years[5]. Taro originated in Southeast Asia and later spread to other parts of the world, including Japan, the Pacific Islands, and Africa. It is a staple crop in the Pacific, Asia, West Africa, and the Americas, feeding over 500 million people worldwide.

Notes

1. *The world's largest plant is an Australian seagrass clone*. (2022, June 1). Natural History Museum. https://www.nhm.ac.uk/discover/news/2022/june/worlds-largest-plant-australian-seagrass-clone.html

2. *Methuselah, a Bristlecone Pine is Thought to be the Oldest Living Organism on Earth*. (2011, April 21). USDA. https://www.usda.gov/media/blog/2011/04/21/methuselah-bristlecone-pine-thought-be-oldest-living-organism-earth

3. Gruber, K. (2022, February 25). *Trees that can walk up to 20m per year*. https://www.bbc.com/travel/article/20151207-ecuadors-mysterious-walking-trees

4. *Heaviest pumpkin*. (2023, October 12). Guinness World Records. https://www.guinnessworldrecords.com/world-records/heaviest-pumpkin

5. Ferdaus, M. J., Chukwu-Munsen, E., Foguel, A., & Da Silva, R. C. (2023). Taro roots: an underexploited root crop. *Nutrients, 15*(15), 3337. https://doi.org/10.3390/nu15153337

Chapter 3

Extreme Natural Phenomena

3.1 The Fastest Winds

- The fastest non-tornado wind speed ever recorded by an anemometer was 253 miles per hour (408 km/h), measured on April 10, 1996, on Barrow Island, Australia, during Tropical Cyclone Olivia. The gust was produced by a smaller-scale circulation within Olivia's eyewall called a mesovortex. However, the fastest wind speed ever recorded at a staffed weather station was 231 mph (372 km/h), measured at Mount Washington Observatory in New Hampshire on April 12, 1934. Mount Washington also holds the record for the highest wind speed in the Northern and Western Hemispheres.
- The strongest tornado. The 1999 Bridge Creek–Moore tornado was an F5 tornado with winds reaching a record-breaking 321 miles per hour (517 km/h). It was one of 58 tornadoes that struck Oklahoma City on May 3, 1999, and was the most destructive of the outbreak. This was the first F5 tornado to hit the Oklahoma City metro area. The tornado tracked for nearly an hour and a half along a 38-mile path, killing at least 36 people and injuring 583 others. It caused $1 billion in damage, equivalent to about $1.8 billion today.

3.2 The Windiest Places

- Commonwealth Bay in Antarctica is known for being the windiest place on Earth.

It's renowned for its katabatic winds, which are caused by cold air flowing down from the Antarctic ice sheet. With an average annual wind speed of 50 mph, gusts in the bay can reach speeds of over 150 mph and can continue for days at a time. The bay's half-moon shape funnels air streams, forcing them to move faster. The lack of landmass to slow down the wind contributes to Antarctica being one of the windiest places on the planet. Commonwealth Bay is also known for being a base for the Australasian Antarctic Expedition.

- Wellington, New Zealand is considered the windiest city in the world. The wind isn't exceptionally strong, averaging 22.7 mph (36.5 kph) year-round, but it experiences gusts over 37 mph (60 kph) for 173 days a year. The strongest gust ever recorded in Wellington, on Mount Kaukau, reached 125 mph (201 kph). Wellington's location on the Cook Strait, between the Pacific Ocean and the South Island of New Zealand, makes it vulnerable to strong winds. The Cook Strait acts like a wind tunnel, funneling ocean breezes into the city. Wellington is nicknamed "Windy Wellington," and its waterfront features a statue called "Solace in the Wind," depicting a person leaning into the breeze. Because of its strong winds, Wellington harnesses this natural resource to generate clean energy.

3.3 The Heaviest Rainfall

- The village of Mawsynram, located in the Indian state of Meghalaya, holds the title for the wettest place on Earth, with an average annual rainfall of 467.4 inches (11,872 mm). This staggering amount is ten times India's national average of 42.6 inches (1,083 mm). In 1985, Mawsynram earned a place in the Guinness World Records for the highest annual rainfall ever documented, recording a mind-boggling 1,042.2 inches (26,471 mm) between 1860 and 1861. Nestled in the Khasi Hills, Mawsynram owes its extraordinary rainfall to the warm, moist air sweeping in from the Bay of Bengal during the monsoon season. In comparison, Cardiff, the rainiest city in the UK, receives an average of just 45.4 inches (1,152 mm) of rain annually.
- In Colombia, López de Micay and Lloró recorded estimated annual rainfall totals of 507.6 inches (12,892.4 mm) and 500.7 inches (12,717 mm), respectively. However, these figures are not widely recognized due to gaps in the available records.

3.4 The Heaviest Snowfalls

- Sukayu Onsen, meaning "acidic hot spring," is one of the snowiest inhabited places in the world. It is located in the city of Aomori in the northern part of Japan's main island of Honshu. It receives an average of 57.9 feet (17.6 meters) of snow annually—the highest in the world for an official reporting station. Its record winter snowfall is 78 feet (23.7 meters). On February 26, 2013, Sukayu Onsen recorded 18.57 feet (5.66 meters) of snow, the highest snow depth ever recorded at a JMA certified weather station. Despite the heavy snowfall, Sukayu Onsen is accessible by road year-round and is famous for its large public hot springs and proximity to skiing. The area's high elevation, cold northern winds, and humid continental climate contribute to its significant snowfall.

- Although Sukayu Onsen is known for its highest annual snowfall, the snowiest place in one year was Paradise, Mount Rainier, Washington. It received 1,224 inches (31.1 meters) of snow from February 19, 1971, to February 18, 1972, which is about the height of a 10-story building. This is also the Guinness World Record for the greatest snowfall in 12 months. Meanwhile, the highest snowfall in a 24-hour period was 100.8 inches (2.56 meters) in Capracotta, Italy, on March 5, 2015. Make sure you stay indoors on those snowy days!

3.5 The Driest Locations

- Arica, Chile, is considered the driest city in the world, with an average annual rainfall of just 0.03 inches (0.08 cm) over a 59-year period. The surrounding Atacama Desert is even drier, with some areas receiving no rain for over 500 years. This aridity results from the Atacama's location in a rain shadow, where mountains block moisture-laden air, allowing only dry air to reach the desert.

- The McMurdo Dry Valleys in Antarctica are another extremely dry place on Earth. Located about 60 miles (97 kilometers) inland from the western coast of McMurdo Sound, they are ice-free and covered in barren rock. This is due to strong, dry katabatic winds with speeds of up to 200 miles per hour (322 km/h), which blow away any snow that falls on the valley floor. Because of their extremely low humidity, both the Atacama Desert in Chile and the McMurdo Dry Valleys are used by NASA to test tools and equipment for Mars missions.

3.6 The Largest Deserts

- It may surprise you, but the largest desert in the world is not the Sahara. Antarctica is the largest desert in the world, covering 5.5 million square miles (14.2 million square kilometers). It's also the coldest and windiest desert on Earth and is home to the geographic South Pole. Antarctica is considered a desert because it receives very little precipitation, with some areas receiving less than 2 inches (51 millimeters) annually. Deserts are defined by low amounts of rain and snow, which lead to little liquid water on the ground, rather than by hot temperatures and sand dunes. In fact, the McMurdo Dry Valleys in Antarctica are one of the driest places on Earth. Despite its harsh conditions, Antarctica is home to a variety of wildlife, including penguins, seals, and orcas.

- The Arctic polar desert is the second-largest desert on Earth, after Antarctica. This vast, cold polar desert encompasses 5.4 million square miles (13.9 million square kilometers) and includes the northernmost regions of the planet, primarily covering parts of Canada, Greenland, and Russia. It's characterized by extremely low temperatures, minimal precipitation, and a landscape devoid of liquid water. The Arctic features large ice sheets, permafrost, and a harsh, windy environment. Despite the challenging conditions, it supports various adapted species like polar bears, walruses, and birds like the Arctic tern and northern fulmar.

- The Sahara Desert is the world's largest hot desert and the third-largest desert overall, after Antarctica and the Arctic. It covers 3.6 million square miles (9.3 million square kilometers), which is comparable in size to China. The Sahara is the hottest desert globally, with an average annual temperature of 86°F (30°C) and a record high of 136°F (58°C). It receives minimal rainfall, with less than 1 inch (25.4 millimeters) annually in many areas. Despite the harsh conditions, the Sahara supports diverse plant life, including cacti, date palms, and spiny plants adapted to minimize moisture loss. It's also home to hundreds of animal species, such as dung beetles, scarab beetles, scorpions, and various ants. The Sahara has a rich history, with human habitation spanning centuries. Notable civilizations like Ancient Egypt and Carthage emerged from this region. Due to its unique landscape, the Sahara has served as a filming location for the Star Wars saga.

3.7 The Coldest Temperatures

- Yakutsk, Russia, is the coldest city in the world as of 2023. As the capital of the Sakha Republic in eastern Siberia, Yakutsk recorded a low temperature of -83.9°F

(-64.4°C) on February 5, 1891. January is typically the coldest month, with an average low of -43.6°F (-42°C). This city of 355,000 is built on permafrost (a layer of frozen ground that includes soil, gravel, and sand), and many houses are constructed on concrete piles. Despite the frigid winters, Yakutsk experiences surprisingly hot summers, with an average high of 78.8°F (26°C) in July and is warmer than London.

- Oymyakon is a remote village in the Sakha Republic of Russia and is located between two valleys. It is known as the coldest inhabited place on Earth, with an average winter temperature of around -58°F (-50°C). In 1924, the temperature plummeted to a bone-chilling -96.2°F (-71.2°C). The village consists of wooden houses built on foundations sunk deep into the permafrost. Residents leave their cars running to prevent them from failing to restart in the extreme cold. Oymyakon once had a population of 2,500, but by 2018, it had dwindled to fewer than 900.

- The coldest place on Earth is the East Antarctic Plateau, where temperatures can plummet to -144°F (-98°C). Located at the center of the continent, the plateau is home to the geographic South Pole. Scientists used satellite sensors to record this extreme temperature on August 10, 2010. The region's clear skies and extremely dry air contribute to these frigid conditions.

3.8 The Hottest Temperatures

- Dallol, Ethiopia, is considered the hottest inhabited place on Earth, with the highest average annual temperature of 93.9°F (34.4°C). However, daytime temperatures can reach over 120°F (49°C). Located in the Danakil Depression, Dallol's extreme environment results from a combination of intense heat, arid conditions, and low elevation.

- The highest air temperature ever recorded on Earth was 134.1°F (56.7°C) at Furnace Creek Ranch in Death Valley, California, on July 10, 1913. In 2024, Death Valley experienced its hottest summer on record, with an average 24-hour temperature of 104.5°F (40.3°C) from June to August. This broke the previous record of 104.2°F (40.1°C) set in 2018 and 2021.

- The Lut Desert in Iran holds the record for the hottest surface temperature on Earth. In 2005, a satellite measured the soil surface temperature in the Lut Desert at 70.7°C (159.3°F). In 2021, a satellite recorded temperatures of up to 81°C (177°F) in the desert. The Lut Desert is located in a basin surrounded by mountains in southeastern Iran. Its location contributes to the extreme dryness, making it inhospitable for most life forms. However, the Lut Desert is home to a variety of resilient animals, including Rüppell's foxes, sand cats, arrow snakes, and golden scorpions.

3.9 The Extreme Earthquakes

- The deepest earthquake ever recorded was a minor aftershock of the 2015 Ogasawara earthquake, occurring at a depth of 751 kilometers (467 miles) below the Earth's surface. This quake was so deep that it originated in the lower mantle, a region where seismologists previously believed earthquakes were impossible. The 2015 Ogasawara earthquake was a magnitude 7.9 quake that shook the Bonin Islands off mainland Japan. The aftershock, too small to be felt at the surface, was detected only by Japan's Hi-net array of seismic stations.

- The strongest earthquake ever recorded was the 9.5 magnitude earthquake in Chile on May 22, 1960. The earthquake occurred at a subduction zone where the Pacific Plate moved under the South American Plate. The earthquake caused tsunamis across the Pacific Ocean, impacting Hawaii, Japan, the Philippines, eastern New Zealand, southeast Australia, and the Aleutian Islands. Hilo, Hawaii, over 10,000 kilometers (6,200 miles) from the epicenter, recorded tsunami waves as high as 35 feet (10.7 meters). The 1960 Chile earthquake led to the development of a global tsunami warning system.

- The 2011 magnitude 9.0 Tohoku earthquake, also known as the Great Sendai Earthquake, was the most powerful earthquake to ever hit Japan. It is likely the most impactful earthquake, moving the main island of Japan by about 8 feet (2.4 meters) and shifting the Earth's axis by approximately 6.5 inches (17 centimeters). This resulted in a 1.8 microsecond shortening of the day's length.

- The 2011 Tohoku earthquake and tsunami is also considered the most expensive earthquake in history, with estimated costs reaching $360 billion. The earthquake destroyed over 123,000 houses and damaged nearly a million more. It also critically damaged backup generators at the Fukushima Daiichi nuclear plant, leading to a nuclear accident that released radioactive material into the atmosphere and ocean.

- Some people may believe the 2004 Indian Ocean earthquake, which claimed the lives of 227,898 people, was the deadliest earthquake, but it was not. The deadliest recorded earthquake in history was the 1556 Shaanxi earthquake in central China, killing an estimated 830,000 people. The earthquake struck a region where most people lived in yaodongs, artificial caves carved from soft rock, which collapsed during the quake. The second deadliest earthquake was the 1976 Tangshan earthquake in China. The magnitude 7.6 earthquake hit the city at 3:42 a.m. on July 28, 1976, when most people were asleep. An estimated 255,000 people died in this earthquake, but some experts believe the toll may have been as high as 655,000.

3.10 The Tallest Waterfalls

- Angel Falls, located in Canaima National Park, Venezuela, is often considered the world's tallest waterfall. With a height of 3,212 feet (979 meters) and an uninterrupted drop of 2,648 feet (807 meters), it is a breathtaking natural wonder. The falls comprise the main plunge, a sloped cascade, and a downstream plunge. Named after American adventurer James Angel, who crash-landed his plane on a nearby mesa in 1937, Angel Falls' status as the tallest waterfall is debated. Some argue that only the uninterrupted drop of 2,648 feet should be considered when measuring a single waterfall's height. If this measurement is used, Tugela Falls in South Africa would claim the title of the world's tallest waterfall.

- Tugela Falls in South Africa stands at 3,110 feet (948 meters) tall, making it the world's second-highest waterfall and the tallest in Africa. It comprises five distinct free-leaping falls, and the highest section measures 1,348 feet (411 meters). The falls is located in the Royal Natal National Park within the Drakensberg mountains, and its top is nearly 10,000 feet (3,048 meters) above sea level. The name Tugela comes from the Zulu word "thukela," meaning "something that startles." If Angel Falls' total height is considered only as its initial 2,648-foot (807-meter) drop, then Tugela Falls would be the world's tallest waterfall.

- However, neither Angel Falls nor Tugela Falls comes close to the Denmark Strait cataract, an underwater waterfall in the Atlantic Ocean between Greenland and Iceland. It is the tallest waterfall of any kind, plunging more than 11,500 feet (3,505 meters) from the Greenland Sea to the Irminger Sea. It carries around 175 million cubic feet (5 million cubic meters) of water per second, equivalent to nearly 2,000 Niagara Falls at peak flow. This underwater overflow is formed by the density difference between water masses on either side of the Denmark Strait.

3.11 The Longest Rivers

- The Amazon River, previously considered the second-longest river after the Nile, is now recognized as the longest river in the world. Measured at least 4,258 miles (6,853 kilometers) long, the Amazon River is not only the longest but also the largest river by size, carrying four times more water than any other river. It accounts for 15-16% of the world's total river discharge into the oceans, pumping 52,834 gallons (200,000 liters) of fresh water into the ocean every second. The Amazon River is surrounded by the Amazon rainforest, which is the world's largest rainforest and houses millions of species. However, the Amazon faces many threats, including deforestation, poorly planned infrastructure development, and climate change.

- The Nile River is about 4,132 miles (6,650 kilometers) long and flows from central Africa to the Mediterranean Sea. It has three main tributaries: the White Nile, the Blue Nile, and the Atbara. The Nile's drainage basin covers about 10% of Africa and has supported agriculture for thousands of years. The Nile was a major trade route for the ancient Egyptians, who used it to transport goods such as gold, ivory, and spices. The Nile is surrounded by many historical sites and landmarks, including the Pyramids of Giza, Luxor Temple, and the Valley of the Kings.

3.12 The Deepest Oceans

- The deepest part of the ocean is the Mariana Trench, which is located in the western Pacific Ocean near the Mariana Islands. The deepest part of the trench is called Challenger Deep and is 36,037 feet (10,984 meters) deep, deeper than Mount Everest is tall. The pressure at the bottom of the Mariana Trench is 1,086 bars, or 15,750 pounds per square inch (psi), which is equivalent to the weight of 100 adult elephants standing on your head. Despite this harsh environment, the trench is home to surprisingly diverse life forms, including the deep-sea crustacean Hirondellea gigas. In 2012, James Cameron made history by becoming the first person to complete a solo submarine dive to Challenger Deep.

- The Puerto Rico Trench is the deepest part of the Atlantic Ocean. Located about 75 miles north of Puerto Rico, the trench reaches a maximum depth of 28,373 feet (8,648 meters). It formed over 70 million years as the North American Plate slid beneath the Caribbean Plate. This geologically active area poses risks of earthquakes and tsunamis but also supports diverse marine life, including a breeding population of humpback whales and other species like Bryde's, sei, fin, and minke whales.

3.13 The Extreme Lakes

- The Caspian Sea situated between Europe and Asia is the world's largest inland body of water. It is bordered by Russia, Azerbaijan, Kazakhstan, Turkmenistan, and Iran and covers approximately 149,200 square miles (386,400 square kilometers), which is larger than Japan. The Caspian Sea stretches nearly 750 miles (1,200 kilometers) in length and has an average width of 200 miles (320 kilometers). Its depth varies considerably, with the shallowest parts in the north averaging 13 to 20 feet (4 to 6 meters) deep, while the southern regions reach depths of 3,360 feet (1,024 meters). Although salty, the Caspian Sea's salinity is only about a third of that found in average seawater. It is fed by the Volga, Ural, and Terek rivers from the north and it supports a diverse ecosystem of fish, mollusks, and other organisms, including species originating from both Arctic and Mediterranean environments.
- Lake Superior is the world's largest freshwater lake by surface area, covering 31,700 square miles (82,103 square kilometers). It is also the world's third-largest lake by volume, holding about 10% of the planet's fresh surface water. With a maximum depth of 1,332 feet (406 meters), Lake Superior is the deepest of the Great Lakes. It is also the coldest of the Great Lakes, with an average water temperature of 40°F (4.4°C). Lake Superior is the habitat of diverse species including mammals, birds, amphibians, and fish.
- Lake Baikal in southern Siberia, Russia is the world's deepest lake, with a maximum depth of 5,387 feet (1,642 meters). It is also the world's largest freshwater lake by volume, containing 20–23% of Earth's unfrozen surface freshwater. Lake Baikal is the world's oldest freshwater lake with an estimated age of 20–25 million years. It extends 395 miles (636 kilometers) in length and 49 miles (79 kilometers) in width, with a surface area of 12,248 square miles (31,722 square kilometers). In comparison, Crater Lake in Oregon is the deepest lake in the United States, with a depth of 1,943 feet (592 meters). Crater Lake is situated within Crater Lake National Park and is known for its mountains, peaks, and evergreen forests.

3.14 The Largest Volcanoes

- Still thinking about Yellowstone's supervolcano? There's one much bigger. Tamu Massif is a massive underwater volcano and the largest volcano on Earth. Located about 1,000 miles (1,609 kilometers) east of Japan in the Pacific Ocean, it is part of the Shatsky Rise underwater mountain range, which formed about 145 million years ago. Tamu Massif covers an area of about 120,000 square miles (310,799 square kilometers), which is roughly the size of New Mexico. Its tallest point is about 6,500 feet (1,981 meters) below the ocean's surface, and its base extends down to about 4 miles (6.4 kilometers) deep. Tamu Massif is about half the size of Olympus Mons on Mars, which is the largest volcano in the solar system and is visible from Earth on a clear night.

- Mauna Loa is the largest active volcano on Earth, covering 2,035 square miles (5,271 square kilometers) and comprising about half of Hawaii Island. Its summit reaches 13,680 feet (4,170 meters) above sea level, but its submarine flanks descend an additional 16,400 feet (5,000 meters) to the ocean floor. This makes Mauna Loa more than 56,000 feet (17,000 meters) tall from base to summit, so massive that its weight bends the ocean floor beneath it! Mauna Loa has erupted 33 times since 1843. Its most recent eruption occurred from November 27 to December 13, 2022.

- However, the gigantic volcanoes mentioned above are not the worst troublemakers. The deadliest volcanic eruption in recorded history was Mount Tambora's eruption in Indonesia on April 10–11, 1815, which killed an estimated 92,000 people. The eruption directly killed an estimated 10,000 people, and an additional 82,000 died from famine and disease caused by the eruption's ash deposits. The powerful eruption expelled 13.2 trillion gallons (50 cubic kilometers) of magma and was heard over 1,200 miles (1,931 kilometers) away on Sumatra Island. The eruption left a collapsed caldera 3.7 by 4.3 miles (6 by 7 kilometers) wide and 3,281 feet (1,000 meters) deep in place of the volcano's summit cone. It is ranked as a 7 on the Volcanic Explosivity Index (VEI), and it is the second-highest rating and considered "super-colossal."

3.15 The Longest Valleys

- The Great Rift Valley is the longest rift valley on Earth's surface, stretching over 4,000 miles (6,400 kilometers) from Jordan to Mozambique in East Africa. This Y-shaped valley comprises a system of interconnected rift valleys, including the

East African Rift, the Red Sea Rift, and the Gulf of Aden Rift. It has an average depth of 2,000–3,000 feet (600–900 meters), though it reaches depths of up to 9,000 feet (2,700 meters) in some places, such as the Mau Escarpment in Kenya. The Great Rift Valley is still active, slowly widening and deepening. It is home to many lakes, volcanoes, deserts, plains, and escarpments, which provide habitat for diverse wildlife, including birds, lions, leopards, elephants, rhinos, buffaloes, cheetahs, giraffes, antelopes, zebras, and primates.

- The Missouri River Valley is the longest river valley in North America. Its river originates in the Rocky Mountains' Bitterroot Range in southwestern Montana and flows east and south for 2,341 miles (3,767 km) before joining the Mississippi River north of St. Louis, Missouri. The valley, which includes the river's floodplains and river bottoms, drains one-sixth of the United States. The Missouri River, often called the "Big Muddy" due to its high sediment load, contributes to the rich alluvial soil along the valley. It was a crucial pathway for westward expansion in the United States and has been a vital artery for transportation, food, trade, and exploration for millions of people over thousands of years.

- However, the two valleys are not even close to the Mid-Atlantic Ridge (MAR), a massive underwater mountain range stretching roughly 16,000 kilometers (10,000 miles). This volcanic mountain range curves from the Arctic Ocean to near the southern tip of Africa. The southern part of the MAR separates the African Plate and the South American Plate. The MAR is a site of seafloor spreading, where magma from the mantle rises to the seafloor, cools, and pushes away from the ridge. This process is responsible for the widening of the Atlantic Ocean at a rate of about 1–10 cm per year.

3.16 The Extraordinary Flatlands

- The West Siberian Plain in central Russia is the world's largest flatlands, covering about 1.2 million square miles (3.1 million square kilometers). It's considered the world's largest unbroken lowland, with more than half of its area less than 328 feet (100 meters) above sea level. The plain is bordered by the Ural Mountains to the west, the Yenisey River to the east, the Kara Sea to the north, and the Torghay Plateau, Kazakh Uplands, and Altai Mountains to the south. Drained by the Ob, Irtysh, and Yenisey rivers and their tributaries, the plain encompasses tundra, forest, and steppe biomes, featuring deep green forests, blue-black pools and ponds, and permafrost. The West Siberian Plain experiences a seasonal temperature differential of up to 158°F (70°C) or more. In the summer, temperatures can reach 86°F (30°C), while in the winter they can plummet to -49°F (-45°C).

- The Tibetan Plateau, also known as the Qinghai-Tibet Plateau or the Qing-Zang Plateau, is the world's largest and highest plateau. It covers more than 0.97 million square miles (2.5 million square kilometers), roughly five times the size of California. It is located in Central Asia and stretches through Tibet, China, and India. The Tibetan Plateau has an average elevation of 14,800 feet (4,500 meters) above sea level and is known as "the roof of the world". The plateau is surrounded by tall mountain ranges, including the Himalayas, and is home to the world's two highest summits, Mount Everest and K2. The Tibetan Plateau formed about 55 million years ago due to the collision of the Indian and Eurasian tectonic plates, and the formation lasted millions of years. This ongoing process continues to push the land upward. The Tibetan Plateau is known for its harsh, cold winters and low oxygen levels. Temperatures can plummet to −40°F (−40°C) in winter, and the air at this elevation contains only about 60% of the oxygen found at sea level.

3.17 The Amazing Basins

- The Pacific is the largest and deepest ocean basin on Earth. It covers more than 59 million square miles (155 million square kilometers), which is over 30% of the planet's surface. It contains over half of the world's free water and is so vast that all of the world's continents could fit inside it. The Pacific Ocean is also the oldest basin, with some of its rocks estimated to be 200 million years old. The Pacific is known as the "Ring of Fire" due to intense volcanic and earthquake activity in the region where the Pacific Plate meets many other tectonic plates. This region is home to 75% of the world's volcanoes and 90% of its earthquakes. The Pacific supports the most diverse marine life of any ocean, including kelp beds, coral reefs, whales, sea turtles, and many fish species.

- The largest river basin in the world is the Amazon River Basin, which covers 2.4 million square miles (6.2 million square kilometers). It is located in northern South America and covers almost 40% of the continent. The Amazon River Basin discharges more than twice as much water as the world's next four largest river basins combined. The Amazon Basin is home to about a third of the world's known species, including 40,000 plant species and 3,000 freshwater fish species. It is also home to more than 400 Indigenous groups who speak around 300 Indigenous languages.

- The largest drainage basin in the United States is the Mississippi/Atchafalaya River Basin (MARB), which is the world's third-largest drainage basin after the Amazon and Congo basins. The Mississippi River drains an area of about 1.2 million square miles (3.2 million square kilometers), covering 41% of the contiguous United States (including parts of 31 US states) and 15% of North America. The Mississippi River is home to more than 325 bird species and 25% of all North American fish species. The basin houses many animals, including endangered species like the Florida panther and peregrine falcon. The basin acts as a flood-relief outlet, slowing the flow of the Mississippi River and protecting communities from storm surges.

3.18 The Vital Rainforests

- The Amazon rainforest is the world's largest rainforest, covering 2.59 million square miles (6.7 million square kilometers), which is about 70% of the size of the continental United States and twice the size of India. The Amazon rainforest spans nine South American countries, including Brazil, Peru, and Colombia, with 60% of it located in Brazil. The Amazon rainforest is home to at least 10% of the world's known biodiversity, including more than 1,800 bird species and many endangered species such as jaguars, harpy eagles, pink river dolphins, and thousands of birds and butterflies. The Amazon helps stabilize the climate by removing 1.2 billion metric tons of carbon from the atmosphere annually, and its forests and soils store an estimated 150–200 billion tons of carbon. The Amazon's thick canopy blocks sunlight, making the forest floor almost black in many parts. When it rains, it can take up to 10 minutes for the water to reach the ground.

- The Congo rainforest covers about 1.2 million square miles (3.1 million square kilometers), making it the world's second-largest tropical rainforest, roughly the size of India. It's located in western equatorial Africa and spans six countries: Cameroon, the Central African Republic, the Democratic Republic of the Congo, the Republic of the Congo, Equatorial Guinea, and Gabon. The Congo rainforest supports vast biodiversity, including over 10,000 plant species and 10,000 animal species. It provides habitat for endangered species like gorillas, elephants, and buffaloes, and is a source of plants used in traditional medicine, including cinchona and rauwolfia. The Congo rainforest is a vital ecosystem, sometimes called the "lungs of Africa" and the "second lung of Earth." The Congo Basin absorbs nearly 1.5 billion tons of carbon dioxide annually, making it the world's largest carbon sink, absorbing more carbon than the Amazon. However, it is threatened by deforestation, and half a million hectares lost in 2022.

3.19 The Longest Lightning Strikes

- The longest lightning strike ever recorded was 477 miles long and stretched from Texas to Mississippi on April 29, 2020. The strike was captured by NOAA satellites and certified by the World Meteorological Organization (WMO). This broke the previous record of 440.6 miles set in 2018 over southern Brazil. While lightning bolts typically measure 2 to 3 miles long, some can stretch across hundreds of miles. Lightning specialist Ron Holle notes that these exceptionally large and long-duration lightning events are not isolated occurrences but happen during active thunderstorms.

- The longest-duration lightning strike lasted 17.2 seconds and occurred on June 18, 2020 over the Argentina-Uruguay border. The strike was recorded by lightning detection equipment on satellites in Earth orbit. Megaflashes are enormous lightning strikes that can last more than 10 seconds and travel for hundreds of miles. They are associated with mesoscale convective systems (MCSs), which are clusters of thunderstorms that can cover several states.

3.20 The Most Frequent Thunderstorms

- The Catatumbo lightning storm in Venezuela is the most intense lightning phenomenon in the world. The lightning occurs 140–160 nights a year and sometimes up to 300 nights a year, with flashes from 16 to 40 times per minute. The lightning density is the highest in the world, with 250 flashes per square kilometer. The lightning occurs over the mouth of the Catatumbo River where it flows into Lake Maracaibo. It is caused by a combination of warm air from the Caribbean Sea and cold air from the Andes Mountains. The name Catatumbo comes from the Bari language and means "House of Thunder." Local fishermen can be struck by lightning if they are out in their boats. According to local news, three or four people die yearly in the lightning capital of the world.

- The village of Kifuka in the Democratic Republic of the Congo (DRC) experiences 158 lightning bolts per square kilometer annually, making it the place with the second-most frequent lightning strikes in the world. Kifuka is located in the mountains of Congo at an elevation of around 5,600 feet (1,700 meters). Its high lightning density is caused by warm Atlantic air colliding with cool mountain air, generating static charges that result in lightning. The lightning is most intense from October to March and less intense from June to August.

3.21 The Highest Tides

- The Bay of Fundy in Canada has the highest tides in the world, with a maximum tidal range of 52 feet (16 meters) and an average tidal range of 38.4 feet (11.7 meters). This far surpasses the average tidal range of 3.3 feet (1 meter) worldwide. The Bay of Fundy's tides are so powerful that about 100 billion tonnes of water flow in and out of the bay during a 12-hour tidal cycle. This is more than the combined flow of all the world's rivers over the same period. The bay's unique shape and resonance, or seiche, are responsible for its extreme tides. The tides vary depending on the positions of the sun and moon, as well as atmospheric conditions. Ungava Bay in Canada has a tidal range of 32 feet (9.8 meters), the second-highest in the world, and Bristol Channel in England and Wales has a tidal range of 31.5 feet (9.6 meters), the third-highest.

3.22 The Largest Waves

- The wave at Lituya Bay in Alaska in 1958 was the world's largest recorded tsunami, reaching a maximum height of 1,720 feet (524 meters). It was caused by a landslide triggered by a magnitude 7.7 earthquake on the Fairweather Fault in southeast Alaska. The wave's impact was devastating, clearing trees to a height of 1,719 feet (524 meters) and sweeping away around 2 square miles (5.2 square kilometers) of forest. This tsunami caused damage at higher elevations than any other recorded, snapping trees and scouring away almost all vegetation. Five people were killed. Lituya Bay is also known for its large tides with a range of about 9.8 feet (3 meters), and its entrance is considered dangerous for navigation, especially when the tidal currents are running.

3.23 The Most Dangerous Places

- Snake Island, also known as Ilha da Queimada Grande, is a small island off the coast of Brazil that is home to thousands of venomous golden lancehead pit vipers. It is so dangerous that only the Brazilian navy and scientists with special permits are allowed to visit. Researchers estimate that there are between 2,000 and 4,000 golden lancehead vipers living on Snake Island, with a density of roughly one to five snakes per square meter. The golden lancehead is one of the most venomous snakes in the world, and it is responsible for a reported 90% of snakebite-related deaths in Brazil. Its venom is fast-acting and can cause kidney failure, internal bleeding, and tissue death. The snakes were trapped on the island around 11,000 years ago when rising ocean levels disconnected it from the mainland. While incredibly dangerous, the golden lancehead's venom has shown promise in the treatment of heart disease, circulatory issues, and blood clots. The island is also a target for wildlife smugglers, known as biopirates, who illegally capture and sell the snakes.

- Fraser Island, also known as K'gari in Australia, is considered one of the most dangerous beaches in the world. It is home to a variety of dangerous wildlife, including dingoes, sharks, and venomous snakes. Dingoes can be unpredictable and dangerous if approached or provoked. Great white sharks and bull sharks can be found in the shallow waters near the shore. The ocean around Fraser Island has frequent riptides, which can be dangerous even for experienced swimmers. Additionally, Fraser Island is very remote, with no lifeguards patrolling the beaches or medical facilities available in case of an accident. However, the island has over 40 lakes and creeks where swimming is considered safe.

- The most dangerous coast. The Skeleton Coast in Namibia is a remote and inhospitable region with shipwrecks, treacherous currents, and sharks. As the world's largest ship graveyard, it is home to thousands of shipwrecks, caused by dense fog, unpredictable currents, shallow sandbars, and rough surf. The Skeleton Coast also hosts 11 species of sharks, making it the third most shark-infested area in the world. Known for its perilous conditions, the San people named it "The Land God Made in Anger." Its remoteness and inaccessibility add to its harshness, with the northern part of the coast closed to normal visitors.

- The most shark-infested beach. Known as the "shark attack capital of the world," Florida has the highest rate of shark attacks globally. New Smyrna Beach in Volusia County, Florida, has had the most shark attacks in the United States, with 303 unprovoked attacks since 1882. The International Shark Attack File (ISAF) reported 259 non-fatal shark bites in the state between 2012 and 2021. Out of these bites,

32 attacks occurred on New Smyrna Beach. According to a Travel Lens study, this number is twice the number at any other U.S. beach.

- The most toxic desert. The Danakil Desert in East Africa is known as the "Gateway to Hell" because of its highly toxic gas emissions. The desert contains active volcanoes and geysers that release toxic gases, including significant amounts of sulfur dioxide, chlorine vapor, and other volcanic gases. These gases are very dangerous to inhale without proper protection; people visiting the area often need to wear gas masks. In fact, the toxic gases can even be fatal to birds that attempt to stop for a drink of water. In addition to the toxic gases, the area experiences extremely high temperatures, making it nearly unbearable for humans. The saltwater reaching the surface can reach 212°F (100°C). The desert is located across Ethiopia, Eritrea, and Djibouti, and is also dotted with extremely acidic and saline hot springs that release additional toxic vapors.

- The North Anatolian Fault (NAF) in Turkey is considered the region with the highest risk of deadly earthquakes. The NAF stretches 932 miles (1,500 kilometers) from the East Anatolian Fault in eastern Turkey to the Aegean Sea and separates the Eurasian plate from the Anatolian plate. It passes within 12 miles (20 kilometers) of Istanbul, Turkey's most populated city that is home to 12.8 million people. The NAF has been the site of some of the most destructive earthquakes in history, including the magnitude 7.9 Erzincan earthquake in 1939, which killed more than 32,000 people, and the magnitude 7.6 Golcuk-Kocaeli earthquake in 1999, which killed over 17,000. The 2023 earthquake, which resulted in more than 55,000 fatalities, occurred on the East Anatolian Fault. Scientists estimate that the Istanbul-Marmara region has a 62% chance of experiencing an earthquake of magnitude 7.0 or greater.

3.24 The Longest Natural Hiking Trails

- The Trans Canada Trail is the world's longest multi-use trail, connecting the Atlantic, Pacific, and Arctic oceans across Canada. The trail spans 17,398 miles (28,000 kilometers) and stretches from coast to coast. It offers a wide range of activities, including hiking, cycling, paddling, horseback riding, cross-country skiing, and snowmobiling. The trail passes through diverse landscapes, including urban, rural, and wilderness areas, connecting 15,000 communities across Canada. Used by more than 2.6 million people each year for physical activity, the trail contributes to physical and mental health. Hiking the entire trail takes about 26 months.

- The Great Western Loop (GWL) is a 6,875-mile (11,064-kilometer) trail notable for its length, diversity, and challenges. As the longest continuous thru-hike in the US, and the world's second-longest trail, it encompasses deserts, mountains, plains, bushwhacking, and trailless routes. The GWL traverses 12 National Parks and over 75 wilderness areas, connecting five existing long-distance trails: the Pacific Crest Trail, Pacific Northwest Trail, Continental Divide Trail, Grand Enchantment Trail, and Arizona Trail. Due to the demanding nature of this hike, timing is crucial for each section. Ryan Jordan and Andrew Skurka created the GWL and completed the inaugural thru-hike in 2007. Skurka's accomplishment earned him the National Geographic Adventurer of the Year award. Hiking the entire GWL takes approximately one year.

- The American Discovery Trail (ADT) is the first and only non-motorized coast-to-coast trail in the United States. It splits into two routes in the Midwest. The northern route, passing through Chicago, covers 4,834 miles (7,780 km), while the southern route, passing through St. Louis, covers 5,057 miles (8,138 km). Combined, the ADT would be about 6,800 miles long, making it the third-longest trail in the world and the second-longest in the US. The ADT connects to many other trails, including the Appalachian Trail, Pacific Crest Trail, and 50 other national scenic and historic trails. It passes through diverse terrain, including cities, towns, forests, mountains, and deserts. The ADT is still under development, and the ultimate goal is to create a completely off-road trail. Hiking the entire ADT takes about a year.

Chapter 4

Unexplained Natural Events and Mysteries

4.1 The Taos Hum: A Sound That Drives People Crazy

- The Taos Hum is a mysterious sound that residents of Taos, New Mexico, have reported hearing since the early 1990s. People who hear the Taos Hum describe it in various ways, including a low rumble or humming sound similar to a diesel engine idling, a faint hissing or buzzing sound similar to a cicada, or a whirring or humming sound. Some also describe it as a pulsating noise, a humming sound similar to a transformer, a droning sound similar to a propeller plane, and a chugging sound similar to a fishing boat. Some say the sound is louder indoors, especially at night, and that it seems to come from nowhere. Others say the sound doesn't change pitch and is only heard in quiet surroundings. The hum is reportedly associated with dizziness, headaches, nausea, nosebleeds, and diarrhea.
- The origin of the Taos Hum remains unknown. Scientists have proposed many theories, but none of them can fully explain the hum. Some attribute it to tinnitus or other biological auditory effects. Others believe the hum is caused by seismic activity or nearby U.S. military testing. Another theory is that the hum may originate from local mechanical sources, such as industrial plants. Some believe the hum is a form of mass psychogenic illness, also known as mass hysteria. Local legends attribute the hum to more sinister sources, such as stoned hippies, secret government mind control experiments, or underground UFO bases. A group of experts from the University of New Mexico, Los Alamos National Laboratory,

Philips Air Force Laboratory, and Sandia National Laboratories investigated the Taos Hum in 1993, but their findings were inconclusive.

4.2 The Bermuda Triangle: Mysterious Disappearances of Ships and Planes

- The Bermuda Triangle is a loosely defined, roughly triangular region in the western North Atlantic Ocean. Its exact boundaries are not universally agreed upon, but it is generally thought to be bounded by the southeastern coast of the U.S., Bermuda, and the Greater Antilles islands. The area is known for the mysterious disappearance of ships and planes. Some famous aircraft and ships that have gone missing in the Bermuda Triangle include:

- The disappearance of Flight 19. On December 5, 1945, five US Navy TBM Avenger torpedo bombers carrying 14 crew members took off from Naval Air Station Fort Lauderdale, Florida, for a training flight to the Bahamas. The flight's mission was to practice bombing runs on a sunken ship. After completing the bombing runs, the compasses malfunctioned, and the crew lost their bearings. The fleet spent the next 3.5 hours searching for the coastline before contact was lost. A Martin PBM Mariner flying boat was immediately dispatched to search for the missing flight, but it, too, vanished along with its 13 crew members. The Navy's official report concluded that the flight's leader, Lieutenant Charles C. Taylor, mistook small islands for the Florida Keys after his compasses malfunctioned. The cause of the compass malfunction remains unknown.

- The disappearance of USS Cyclops. The USS Cyclops was a World War I ship and the largest in the Navy at the time. This giant vessel carried 306 crew and passengers, along with a heavy cargo of manganese ore. It vanished on March 3, 1918, while traveling from Barbados to Baltimore. Expected to arrive in Baltimore on March 13, 1918, the ship never reached its destination. Its last known location was Barbados, and it vanished without a trace, with its last known message stating "Weather Fair, All Well." The ship did not send out an SOS before vanishing, indicating no apparent distress, and no further communication was ever received. The wreck of the Cyclops has never been found, and on June 1, 1918, Assistant Secretary of the Navy Franklin D. Roosevelt declared the ship lost with all hands deceased. Some claimed that a German U-boat or raider attack sank the ship, but German authorities at the time denied any knowledge of it. Others suspect that the Bermuda Triangle, structural failures due to the ship's heavy cargo, or an unexplained explosion was responsible. The fate of the Cyclops and the cause of its sinking remain a complete mystery.

- The disappearance of Witchcraft. On December 22, 1967, a 23-foot cabin cruiser named *Witchcraft* mysteriously disappeared off the coast of Florida. Its owner, Daniel S. Burack, was a hotel magnate and seasoned yachtsman, and he had invited Rev. Padraig Horgan to join him on a short cruise to view Christmas lights along the Florida coast. Around 9 p.m., Burack sent out a distress signal, reporting that the *Witchcraft* had struck something underwater but was not in immediate danger of sinking. The Coast Guard responded within 19 minutes to the boat's last reported location near Buoy No. 7 in Miami Harbor, but the vessel, Burack, and Horgan had vanished. Despite an extensive search covering 24,500 square miles, including areas of the Bermuda Triangle, no trace of the boat or its passengers was ever found. The *Witchcraft* was equipped with flotation devices that should have prevented it from sinking, adding to the mystery of its disappearance. This event remains unsolved decades later. Some theorize that vortices sucking objects into other dimensions, oceanic flatulence, or disruptions in geomagnetic lines of flux were responsible for the ship's disappearance, while others claim extraterrestrials took the ship into a different dimension of space. The exact cause of Witchcraft's disappearance remains a mystery.

- However, despite these well-known incidents, other people believe the connection between the Bermuda Triangle and mysterious disappearances is an urban legend. People navigate the area every day without incident, and there's no evidence that mysterious disappearances occur more frequently there than in other well-traveled areas of the ocean. What is your opinion?

4.3 Ball Lightning: A Mysterious Floating Light

- Ball lightning is a rare, mysterious, and poorly understood natural phenomenon. It is often described as a luminous ball that floats above the ground or drifts erratically, sometimes leaving smoky trails. Ball lightning typically lasts only a few seconds and is often accompanied by a hissing sound and a distinct odor. It has been reported by thousands of people, including peasants in the Middle Ages and Tsar Nicholas II, and has even been known to cause damage through burning or melting. Some famous cases of ball lightning includes:

- Georg Richmann's death. Georg Wilhelm Richmann, a Baltic German physicist, was killed by ball lightning while conducting an electrical experiment on August 6, 1753. He was attempting to measure the response of an insulated rod to a nearby storm. During the thunderstorm, a ball of lightning jumped from the rod and struck Richmann's forehead, killing him instantly. The incident also caused an explosion that knocked out Richmann's engraver, split the door frame, and tore the door off its hinges. Richmann's death is believed to be the first recorded fatality from ball lightning and the first known instance of someone dying while conducting electrical experiments.

- One day in June 1841, at 6 p.m. local time, Lorenzo Butti, a marine painter for the Empress of Austria, was alarmed by terrified calls from the Corsa dei Servi (central street, now Corso Vittorio Emanuele II) in the center of Milan, Italy. There was a heavy thunderstorm, and Butti looked out the window and saw people running in the rain, fleeing a reddish-yellow ball of lightning. It flew at the height of the second floor, rose higher, and then crashed into a church tower cross and exploded. Butti wrote about his experience to the physicist Arago.

- In the 1960s, the crew of a C-133A cargo plane flying between California and Hawaii witnessed a ball lightning event. It was preceded by a glow of St. Elmo's Fire, an electrical discharge caused by static electricity from the plane's movement through the air. The ball lightning formed over the central metal divider between the glass panes of the pilot's window and slowly floated downward into the cockpit. It then turned left toward the crew lounge doorway before turning right toward the Loadmaster's seat. All five crew members witnessed the event and confirmed seeing the same thing. The Engineer and Loadmaster searched the plane but found no damage. The flight was in a fog-like layer of cloud, with no turbulence. No lightning was seen outside the aircraft, and the radar showed no clouds.

- Scientists have yet to successfully recreate ball lightning in laboratory conditions. Some prevailing theories include the silicon hypothesis, which posits that when lightning strikes silicon-rich soil, a chemical reaction occurs and creates a glowing ball. Alternatively, the Maser-Soliton Theory hypothesizes that outdoor ball lightning is caused by an atmospheric maser, a phenomenon similar to a laser but operating at a much lower energy level. The exact cause of ball lightning remains a mystery.

4.4 The Tunguska Event: The Largest Impact Recorded in History

- On June 30, 1908, a massive explosion occurred in Siberia, Russia. The explosion was estimated to be equivalent to 10–40 megatons of TNT, which is thousands of times more powerful than the Hiroshima atomic bomb. The blast occurred in the East Siberian taiga, near the Podkamennaya Tunguska River and flattened an estimated 80 million trees over 830 square miles (2,150 square kilometers) of forest. It generated a powerful shockwave heard up to 620 miles (1,000 kilometers) away and registered seismic shockwaves as far away as England. Additionally, the explosion created a bright glow that lit up the sky over Siberia and parts of Europe. Eyewitnesses reported seeing a blinding light, hearing a thunderous noise, and feeling intense heat that burned their skin. The only remnants of the shattered object were a few tiny fragments, each measuring less than a millimeter across. The cause of the explosion remains unknown, and there is little evidence of its origin. The leading theory points to an asteroid or comet; however, no crater or meteorite remnants have been found. Scientists believe the object exploded at an altitude of 3–6 miles (5–10 kilometers) in the atmosphere, causing an air burst.
- On February 15, 2013, a meteor exploded over Chelyabinsk, Russia, providing researchers with new insights into the 1908 Tunguska event. Using modern computer models, video footage, and damage maps, scientists determined that the Chelyabinsk meteor was a stony asteroid about the size of a five-story building. It broke apart 15 miles (24 kilometers) above the ground, generating a shockwave equivalent to a 550-kiloton explosion. This blast shattered a million windows and injured over a thousand people, but it wasn't powerful enough to topple trees or buildings. Scientists estimate that impacts of this kind occur roughly every 10 to 100 years. The Chelyabinsk event helps us understand what might have happened on the day of the Tunguska explosion.

4.5 Hessdalen Lights: The Most Mysterious Lights on Earth

- The Hessdalen lights are a phenomenon that has been reported in the Hessdalen valley since at least the 1930s. The lights can appear in various colors, including white, yellow, and red. They can appear both above and below the horizon, and can float through and above the valley. The lights may appear for a few seconds to over an hour, and can move at varying speeds, from extremely fast to a slow swaying motion. They can also hover in mid-air. Activity was particularly high between

1981 and 1984, with sightings occurring 15–20 times per week. Since then, the number of sightings has decreased. Several projects have been initiated to investigate the lights, including "Project Hessdalen" and "The Triangle Project". The Hessdalen Automatic Measurement Station (Hessdalen AMS) was established in 1998 to record the lights.

- Despite scientists' investigations, the origin of the mysterious lights remains unknown. One recent study suggests that the phenomenon could be attributed to the combustion of airborne dust from mining activities in the area, as Hessdalen has large deposits of scandium. The analysis identified hydrogen, oxygen, and other elements, including titanium, within the lights. However, the exact mechanism behind this phenomenon remains incompletely understood.

4.6 Lake Anjikuni Disappearance: The Unsolved Vanishing of an Inuit Village

- On a cold November day in 1930, fur trapper Joe Labelle arrived at an Inuit village near Lake Anjikuni, Manitoba, Canada, expecting a warm welcome as a frequent visitor. However, he found the village deserted, with tents still standing, food cooking over fires, and personal items like unfinished clothing left behind, suggesting a hasty departure. Adding to the mystery, a freshly dug grave appeared disturbed, and several sled dogs were found dead. Despite searches by the North-West Mounted Police, no bodies or evidence of the villagers' whereabouts were discovered. Furthermore, despite the sudden disappearance, the search team found no signs of struggle or violence in the village. However, members of the Royal Canadian Mounted Police (RCMP) reported seeing pulsating lights in the sky above Lake Anjikuni during their search, leading UFO enthusiasts to propose that the villagers might have been abducted by aliens. The story gained widespread attention after journalist Emmett E. Kelleher reported it in the Danville Bee newspaper, and it was later featured in books like Frank Edwards' "Stranger than Science." Some people accused Kelleher of embellishing details, and due to a lack of evidence, the RCMP dismissed the case in 1931.

4.7 The Mary Celeste: The Ghost Ship

- The Mary Celeste was a Canadian-built, American-registered ship discovered abandoned in the Atlantic Ocean in 1872. The fate of its 10-person crew who disappeared remains one of the most enduring maritime mysteries. Built in Canada in 1861 and originally named the Amazon, the Mary Celeste was later sold to an American and underwent several structural changes. On November 7, 1872, the Mary Celeste set sail from New York City, bound for Genoa, Italy, carrying 1,701 barrels of alcohol. Captain Benjamin Briggs, his wife Sarah, and their two-year-old daughter Sophia were on board, but their seven-year-old son, Arthur, was left behind in New York. On December 4, 1872, the Canadian brigantine Dei Gratia found the Mary Celeste adrift in the Atlantic Ocean. The ship was undamaged, seaworthy, and still partially under sail. Its cargo of 1,701 barrels of denatured alcohol was intact, but the crew and their lifeboat were missing. The ship's log had last been entered 10 days before its discovery, and the ship was still well-provisioned. Theories about the crew's disappearance include coal dust and construction debris fouling the ship's pumps, or Captain Briggs abandoning ship due to uncertainty about its seaworthiness. Other theories include a false alarm leading to an unnecessary evacuation, an explosion caused by alcohol fumes, a waterspout sweeping the crew overboard, and a pirate attack. Although scientific investigations favor the explosion theory, which would have forced the crew to abandon ship, the mystery of the Mary Celeste is likely never to be fully solved.

4.8 Earthquake Lights: Mysterious Lights Accompanying The Shake

- Earthquake lights, also known as earthquake lightning or earthquake flashes, are luminous optical phenomena that can appear in the sky before, during, or near an earthquake. They may also accompany volcanic eruptions or tectonic stress. Earthquake lights occur shortly before or during a seismic event, as far as 370 miles (600 kilometers) from the quake's epicenter, and on or near rift valleys. They can manifest as balls of light, sheet lightning, streamers, steady glows, or bluish flames. They can also resemble bright lightning strikes emanating from the ground and extending up to 650 feet (200 meters). Earthquake lights are typically white to bluish in color but can encompass a wider range of colors. They can last from a fraction of a second to tens of minutes. They are most common in Italy, Greece, France, Germany, China, and parts of South America. Below are some examples of major earthquakes accompanied by earthquake lights:

- On August 15, 2007, bright flashes of light lit up the sky before the 8.0-mag nitude Pisco earthquake in Peru. Security cameras and seismic records captured the precise timing and location of the light flashes. On April 6, 2009, four-inch flames of light flickered above a stone street before the magnitude 6.3 L'Aquila earthquake in Italy. On September 7, 2021, blue lights, resembling flashes of cerulean lightning, illuminated the night sky during the magnitude-7.0 earthquake in Acapulco, Mexico. On February 6, 2023, multiple lights appeared continuously during the 7.8 magnitude Turkey–Syria earthquake. During 2023's magnitude 6.9 Marrakesh-Safi earthquake in Morocco, blue light flashes were seen throughout the tremor.

- Some propose that earthquake lights are caused by tectonic stress breaking apart oxygen atoms within rocks, which then flow toward the surface and ionize air pockets, forming a charged gas. However, there's no broad consensus on the cause of earthquake lights. Some scientists doubt the validity of any of the reports, while others believe that at least some reports plausibly correspond to earthquake lights. Some reports have even been linked to electrical arcing from shaking power lines.

4.9 The Unexplained Forests

- The Crooked Forest in Poland is a grove of 400 pine trees located in the village of Nowe Czarnowo, near the town of Gryfino in West Pomerania, Poland. The trees are bent at a 90-degree angle at their base, forming a J-shape, before curving back to grow straight upward. Despite their unusual curves, the trees are mostly healthy and grow up to 50 feet (15 meters) tall. The trees are a protected natural monument and a popular tourist attraction. There are many theories about how the trees acquired their unique shape. Some believe that foresters intentionally bent the trees to create curved timber for furniture or boat building. Others propose that heavy snow or gravitational pull may have weighed the trees down as they grew. Some theories even suggest that the trees' growth may be the result of a genetic mutation or a UFO landing. Despite numerous hypotheses, the mystery of the Crooked Forest has never been definitively solved.

- The Dancing Forest in Russia is a pine forest in Kaliningrad Oblast, Russia, which is located on the Kruglaya Dune of the Curonian Spit between the Baltic Sea and the Curonian Lagoon. It is known for its unusually twisted trees, which are bent into shapes like rings, hearts, and spirals. The exact cause of the trees' distortion is unknown, but some possible explanations include caterpillar activity, the area's geomagnetic fields, and human manipulation. Local legend claims the trees were made to dance to prove the power of the Christian God. The Dancing Forest has become a popular source of hope, with some believing that climbing through one of the rings can add a year to a person's life or grant a wish.

Chapter 5

The Fascinating Universe

5.1 The Wonders of Gravity

- Gravity is a fundamental force that shapes the universe. It is the primary force that determines the shape of objects in the universe, pulling matter together to form spheres. This is why planets, stars, and other large objects are round.

- Gravity is the weakest of the four fundamental forces (gravity, electromagnetism, the strong nuclear force, and the weak nuclear force) in the universe. For example, a small magnet can overcome the gravitational pull of the entire Earth to lift a paperclip. However, gravity has an infinite range and plays a crucial role in the formation of the universe's structure.

- While mass, a fundamental property of matter, remains constant regardless of location or gravity, weight can vary under different gravitational forces. For instance, a 100-kilogram calibration weight would have the same mass on both Earth and Mars, but its weight would be 100 kilograms on Earth and only 37.7 kilograms on Mars.

- According to Einstein's theory of general relativity, gravity travels at the speed of light, which is approximately 186,282 miles per second (299,792 kilometers per second). If the Sun were to suddenly vanish, it would take 8 minutes for Earth to feel the loss of its gravitational pull—this is the same amount of time it takes for sunlight to reach us.

- Gravitational waves are ripples in space-time caused by some of the universe's most energetic and violent events, such as colliding black holes, colliding neutron stars, or supernovae (massive stars exploding at the end of their lives). Invisible to the human eye, gravitational waves, like gravity itself, travel at the speed of light (186,282 miles per second, or 299,792 kilometers per second), squeezing and stretching anything in their path as they propagate. They can pass through matter without being scattered or obstructed, carrying valuable information about the objects and events that created them. Scientists can use gravitational waves to study the origins of the universe.

5.2 The Fascinating Nature of Time: Is It An Illusion?

- Although time appears linear, we can actually see light from the past because light takes time to travel to Earth. The farther away an object is, the longer it takes for its light to reach us, meaning the light is older upon arrival. For example, the cosmic microwave background, the farthest light we can see, is the remnant radiation from the Big Bang, which occurred approximately 13.8 billion years ago. We see the Sun as it was about 8 minutes ago because that's how long it takes for sunlight to reach Earth. Similarly, we see Jupiter as it was about 48 minutes ago, and Vega as it was 25 years ago. Stars within 4,000 light-years of us appear as they were up to 4,000 years ago. It's even possible that some stars we see tonight no longer exist, as their light has taken years to reach us.

- An observer in motion experiences time more slowly than a stationary observer. For instance, astronauts on the International Space Station (ISS), which travels at approximately 17,500 miles per hour (28,000 kilometers per hour), age 0.007 seconds slower than people on Earth for every 30 minutes they spend in space. So, if you want to stay young longer, consider interstellar travel!

- If we were to travel to a star 10 light-years away at the speed of light, time would appear to freeze. In other words, the journey would feel almost instantaneous for us, but 10 years would have elapsed on Earth. However, as we approach the speed of light, our mass increases significantly, eventually approaching infinity. Because the energy required to propel us would also become infinite, we can never truly reach the speed of light.

- In 2000, former University of Connecticut physics professor Ronald Mallett proposed that a ring laser could bend spacetime sufficiently to enable time travel.

A ring laser, created by generating a continuous rotating beam of light, could theoretically produce a gravitational field similar to a black hole through its concentrated, rotating light. This field could, in theory, create time loops or "closed timelike curves," allowing for travel back in time. Mallett began testing the idea in 2019 but has yet to prove his hypothesis.

5.3 The Mysterious Black Holes

- Black holes are regions in space where gravity is so strong that it warps space-time, preventing even light from escaping. These incredibly dense objects pack an enormous amount of mass into a tiny volume. Black holes often form when massive stars collapse at the end of their life cycles.

- Black holes range in size from the size of an atom to millions of times the mass of the sun. There are at least three types of black holes: 1) Primordial black holes, which are thought to have formed in the first moments after the Big Bang. They can be as small as an atom but with the mass of a large mountain. 2) Stellar black holes. They formed from the collapse of massive stars, which can be up to 20 times the mass of the sun. 3) Supermassive black holes. They are the largest with millions or billions of times the mass of the sun. They are found at the center of most large galaxies, including our own Milky Way, which harbors a supermassive black hole Sagittarius A* with a mass of about 4.3 million suns. The largest known black hole in the universe is TON 618, weighing around 40 billion times the mass of our sun and located 10 billion light-years from Earth.

- The size of a black hole's event horizon, also known as its Schwarzschild radius, is directly proportional to its mass. If Earth were compressed into a black hole, its event horizon would be roughly the size of a marble (0.2 to 0.8 inches, or 0.5 to 2 centimeters in diameter).

- Black holes can merge to form a single, larger black hole. This process is extremely violent and releases a tremendous amount of energy. The collision creates ripples in the fabric of spacetime, known as gravitational waves. In 2015, scientists, using the LIGO detector, observed the merger of stellar-mass binary black holes for the first time. This merger produced a black hole with a mass approximately 62 times that of the sun.

- If you were to fall into a black hole, you would experience "spaghettification"—your body would be stretched into a long, thin strand due to the immense gravitational pull. The part of your body closer to the black hole would experience a significantly stronger force than the part further away, essentially tearing you apart before you even reach the event horizon (the point of no return). Moreover, due to gravitational time dilation, as you approach the black hole, time would appear to slow down for you relative to an outside observer. To an observer watching you fall into the black hole, you would appear increasingly redshifted and dimmer as you approach the event horizon, eventually fading from sight.

- Scientists estimate that there are around 100 million black holes in the Milky Way galaxy, yet only a few dozen have been identified. This discrepancy arises because black holes are inherently difficult to detect, as they do not emit light. Astronomers can, however, detect black holes by observing the deflection of starlight as it passes near a black hole's gravitational field. The closest known black hole to Earth is Gaia BH1, located 1,560 light-years away.

5.4 Dark Matter And Dark Energy: The Universe Is Mostly Dark To Us

- Dark matter is invisible because it doesn't interact with light or other forms of electromagnetic radiation. This means that it doesn't absorb, reflect, or emit light, making it undetectable (thus "dark") to us. However, dark matter has gravitational forces, and its gravity affects the movement of galaxies and stars; this is how we know it exists.

- Dark matter makes up more than 80% of the total mass of the universe. Although invisible, dark matter acts like a glue, playing a key role in galaxy formation and distribution. Without dark matter, galaxies would fly apart due to the speed at which they are moving.

- Dark matter, unlike regular matter such as stars and planets, isn't composed of atoms. Scientists hypothesize that dark matter consists of subatomic particles yet to be discovered, although its exact composition remains a mystery.

- Dark energy is a mysterious force that drives the accelerated expansion of the universe, pushing galaxies apart and fueling this expansion. It is sometimes referred to as "anti-gravity" because it counters the force of gravity, which pulls galaxies together.

- Dark energy constitutes approximately 68% of the universe's total energy content. It is uniformly distributed throughout the universe, both in space and time. This even distribution means it affects the universe as a whole.

- Although dark energy and dark matter are both invisible, they have different effects on the universe. Dark matter pulls galaxies together, while dark energy pushes them apart.

- Together, dark energy and dark matter make up most of the universe, around 95%. Only about 5% of the universe is the "normal" matter that makes up stars, galaxies, and planets. The universe is mostly dark to us!

5.5 The Fascinating Night Sky: A Closer Look

- There are approximately 9,000 visible stars in the night sky, and on a moonless night with minimal light pollution from cities, the naked eye can see around 5,000 to 6,000 of them. These stars are usually no more than 10,000 light-years away (one light-year is the distance light travels in one year, or 5.879 trillion miles), and most stars are still alive when their light reaches Earth. In comparison, there are approximately 100 billion stars in our Milky Way galaxy, and astronomers estimate that there are approximately 200 billion trillion stars in the observable universe!

- With a small telescope capable of detecting fainter stars (for example, stars of magnitude 10; the larger the number, the fainter the star, and human eyes can see up to magnitude 6), we can see over 140,000 stars in the night sky. Some of these stars are so far from Earth that they might have already died by the time their light reaches us.

- Stars are giant balls of hot gas, mostly hydrogen, with some helium and small amounts of other elements. A star's color indicates its surface temperature, with blue stars being the hottest. As a star's temperature increases, its color changes from red to orange to yellow to white, and finally to blue. Red stars are the coolest, with temperatures of about 3,800°C. Orange stars have temperatures of roughly 4,500°C. Our Sun is a yellow star, with a surface temperature of around 6,000°C. Blue stars have surface temperatures exceeding 10,000°C, and the hottest blue stars, known as blue supergiants, can have temperatures of over 40,000°C.

- Blue stars are also the biggest and most massive stars in the galaxy, making them visible in the sky even when they are far away. One of the most well-known blue supergiants in the night sky is Rigel, the brightest star in the constellation Orion (the "hunter"). Rigel's surface temperature is around 11,600°C, which is much hotter than the Sun. Scientists estimate that Rigel has an incredible 120,000 times the brightness of our Sun, but due to its distance from us (863 light-years away), it is only the seventh-brightest star in the night sky.

- The brightest red giant in the night sky, Arcturus, is located in the constellation Boötes (the "herdsman") and is approximately 37 light-years from Earth. As the fourth-brightest star in the night sky, Arcturus derives its name (which means "keeper of the bear") from its position near the constellation Ursa Major (the "great bear"). Despite having only 1.5 times the mass of the sun, Arcturus shines 113 times brighter and releases 215 times more heat due to its immense size. Arcturus is estimated to be 7 billion years old, and it is approaching the end of its stellar life. It will eventually transform into a white dwarf—a small, incredibly dense star—once it exhausts its nuclear fuel.

- Sirius is the brightest star in the night sky and it is visible from both the Northern and Southern hemispheres. Located in the constellation Canis Major (the "greater dog"), Sirius is also known as the Dog Star or Sirius A. It is approximately 8.6 light-years away from Earth and has an apparent magnitude of -1.46. Its name, derived from Greek, means "glowing." Sirius is most prominent in the winter months in the Northern Hemisphere. Only a few planets, the full moon, and the International Space Station outshine Sirius when they are above the horizon. You can find Sirius by following the line of Orion's (the "hunter") belt stars downward, toward the southeast.

- The second brightest star in the night sky is Canopus. It has a visual magnitude of -0.74 and is located 310 light-years away from Earth. Located in the southern constellation Carina (the "ship's keel"), Canopus is primarily visible in the Southern Hemisphere. Its visibility is limited by latitude, and people living at 37 degrees north or higher, such as most of the United States, Europe, and Canada, cannot see Canopus because it remains below the horizon. In Greek mythology, Canopus was the pilot of King Menelaus of Sparta's ship during the Trojan War, while in Chinese culture, Canopus is associated with longevity.

5.6 Heavenly Visitors: Spectacular Comets and Meteor Showers

Comets

- Comets are celestial bodies that are made of ice, gas, and dust, and have long tails. They are remnants from the formation of our solar system about 4.6 billion years ago, and are sometimes called "dirty snowballs". Comets may have brought water and organic compounds to early Earth and other parts of the solar system, which could be the building blocks of life.

- Based on NASA's estimation in 2023, there are about 3,743 known comets in our solar system. However, astronomers believe that there are billions more comets orbiting the sun in the Kuiper Belt and Oort Cloud, which are located beyond Neptune and Pluto. Halley is the most well-known comet with a regular 76-year journey cycle around the Sun. It was last seen in Earth's skies in 1986 and will return in 2061. Below are the spectacular comets that will be visible in 2024 and 2025.

- One of the most stunning comets visible from Earth in 2024 is C/2023 A3 (Tsuchinshan-ATLAS). This comet reached its perihelion (the closest point to the Sun) on September 27, 2024, with a brightness of magnitude 2.9. By October 12, 2024, it will be at its closest approach to Earth, shining at an impressive magnitude of -4. This will make it comparable to the brightness of Venus! The comet will initially be visible in the morning sky of the Southern Hemisphere with binoculars or the naked eye. It will become a fascinating sight in the Northern Hemisphere after October 14, 2024. The best viewing time will be shortly after sunset when the comet can be seen in the western sky until around October 24, 2024.

- Another bright comet expected to appear in 2024 is A11bP7I, or C/2024 S1 (ATLAS). This comet will reach perihelion on October 28, 2024, and could become brighter than Venus, possibly even visible during the day. It belongs to the Kreutz sungrazers, a group of comets that pass extremely close to the Sun. The best views will be from the Southern Hemisphere in the evening sky. If A11bP7I survives its close encounter with the Sun, it will later be visible in mid-northern latitudes during the morning, offering another chance for observation.

- Later in the year, periodic comet 333P/LINEAR from the Jupiter family will reach its perihelion on November 29, 2024. With a brightness of magnitude 9.8, this comet will be observable through large binoculars or a small telescope (our naked eyes can only see up to magnitude 6 and will not see it). By December 9, 2024, it will reach its closest approach to Earth, brightening slightly. This comet will be visible in both hemispheres, offering amateur astronomers a good opportunity to catch a glimpse.

- On January 13, 2025, comet C/2024 G3 (ATLAS) will reach perihelion with a potential brightness of magnitude -2.0. At its closest approach to Earth, it may reach magnitude -1.8, making it visible to the naked eye even during daylight hours. However, the comet's fate remains uncertain, as it might not survive its journey around the Sun. If it does, it will be visible in the Northern Hemisphere before perihelion and in the Southern Hemisphere after. Its proximity to the Sun in the sky during these times will make it difficult to observe.

Meteor Showers

- Meteor showers occur when Earth passes through the debris left by comets and asteroids. As this debris burns up in Earth's atmosphere, it creates dazzling streaks of light across the sky. Every year, there are over 100 meteor showers, and about 30 of them are visible from Earth. Some are brief and last only a few years, while others happen annually. The most extraordinary meteor showers, known as meteor storms, show up in our skies only once or twice in a lifetime. Below are some of the most spectacular meteor showers.

- The Orionids, a meteor shower radiating from the constellation Orion (the "Hunter"), are considered "one of the most beautiful showers of the year" by NASA. They originate from the debris of Halley's Comet and typically peak in late October. In 2024, the Orionids will peak on the night of October 20-21, with the best viewing time after midnight. Under dark skies, expect to see an average of 20 meteors per hour, with the possibility of up to 50-75 during peak activity.

- The Perseids, a meteor shower radiating from the constellation Perseus (the "Hero"), are caused by Earth passing through the debris left behind by Comet Swift-Tuttle, which is known as the Perseid cloud. First observed around 2,000 years ago, the Perseids are visible in mid-August and peak between August 8 and 14. The best time to see them is in the early morning hours when the moon's light is less intrusive. The Perseids are one of the most plentiful showers, and viewers can expect to see about 50 to 100 meteors per hour during peak activity.

- The Geminids, a meteor shower radiating from the constellation Gemini (the

"Twins"), are one of the most reliable and spectacular annual meteor showers. They are visible every year from about November 19 to December 16, peaking in mid-December, typically around December 13 and 14. First observed in the mid-1800s, the early Geminids only produced a modest 10-20 meteors per hour; however, over time, this number has grown to up to 120 meteors per hour under ideal conditions, making the Geminids a major meteor shower. Unlike most meteor showers that originate from comets, the Geminids come from an asteroid named 3200 Phaethon. Its meteoroids are denser than typical comet dust, which makes Geminids meteors uniquely bright, fast, and distinctly yellow. The best time to watch the Geminids is during the night and pre-dawn hours in mid-December. The shower is visible worldwide, with peak visibility beginning around 9 or 10 p.m. To get the best view, find a location far from city lights and give your eyes about 30 minutes to adjust to the darkness.

- The Leonids, a meteor shower radiating from the constellation Leo (the "Lion"), are an annual meteor shower, though they typically produce a more modest 15 meteors per hour. The Leonids originate from comet 55P/Tempel-Tuttle, which takes 33 years to orbit the Sun. This meteor shower will be active between November 3 and December 2 in 2024, with a peak on November 16. The Leonids are known for their bright, colorful, and fast meteors, which travel at an astonishing speed of 44 miles (71 kilometers) per second! Every 33 years, the Leonids can produce a spectacular meteor storm, with up to thousands of meteors per hour. The most famous storm occurred in 1966 when thousands of meteors per minute were observed, and the last storm took place in 2002. The Leonids are also notable for producing fireballs—explosions of light and color caused by larger particles of cometary material—and Earth-grazer meteors, which skim the horizon, leaving long, colorful trails. The best time to observe the Leonids is after midnight from a dark location. Be sure to look away from the constellation Leo to enjoy longer and more spectacular meteors.

- The Quadrantids, a meteor shower radiating from the old constellation Quadrans Muralis (the "mural quadrant"), now part of the modern constellation Bootes (the "herdsman"), are one of the most impressive annual meteor showers. Like the Geminids, the Quadrantids originate from an asteroid—2003 EH1—rather than a comet. The Quadrantid meteor shower will be active from December 12, 2024, to January 12, 2025, peaking around January 3, 2025. This meteor shower has a notably narrow peak, lasting only about six hours, because Earth crosses its meteor stream at a perpendicular angle. During the peak, under ideal conditions, observers can expect to see 60 to 200 meteors per hour. Like the Leonids, the Quadrantids are also known for producing bright, long-lasting fireballs. The Quadrantids are best viewed from the Northern Hemisphere during the night and pre-dawn hours.

5.7 Between Light and Shadow: Fun Facts About Solar and Lunar Eclipses

Solar Eclipse

- The longest recorded total solar eclipse occurred on June 15, 743 BC, and lasted 7 minutes and 28 seconds. The eclipse was visible in the Indian Ocean, off the coast of Kenya and Somalia. The longest possible duration for a total solar eclipse is 7 minutes and 32 seconds. This theoretical maximum would occur when the sun is at aphelion (its farthest point from Earth) and the moon is at perigee (its closest point to Earth).

- On May 29, 1919, a solar eclipse provided key evidence for Albert Einstein's general theory of relativity. According to the theory, the Sun's massive size creates a curvature in spacetime. When light from distant stars passes near the Sun, it follows this curved spacetime, bending its path. During the 1919 solar eclipse, English astrophysicist Arthur Eddington and his team photographed stars near the Sun (normally invisible due to the Sun's brightness). They found that the stars' apparent positions had shifted, and the amount of deflection precisely matched Einstein's predictions. This confirmed Einstein's general theory of relativity.

- The solar eclipse that occurred on November 3, 2013, was a rare hybrid eclipse. It began as an annular eclipse but transitioned into a total eclipse as the Moon's distance from Earth decreased. A hybrid eclipse occurs when the Moon's distance from Earth changes during the eclipse, causing it to shift between total and annular. The next hybrid eclipse will occur on November 14, 2031, passing primarily over the Pacific Ocean. The next two hybrid eclipses are expected in November 2049 and May 2050.

- The solar eclipse of August 21, 2017, was a total solar eclipse visible across the entire contiguous United States for the first time since 1918. Also known as the "Great American Eclipse," its 70-mile-wide (110 km) path of totality (where the

Moon completely blocks the Sun's light) spanned 14 states, including Oregon, Idaho, Montana, Wyoming, Nebraska, Kansas, Iowa, Missouri, Illinois, Kentucky, Tennessee, Georgia, North Carolina, and South Carolina. The total eclipse was first visible in Oregon at 10:15 AM PDT and ended in Charleston, South Carolina, at around 2:41 PM EST.

Lunar Eclipse

- Have you ever seen a blood moon? It's a moon with a spooky, copper-red color. This phenomenon occurs during a total lunar eclipse, when the Earth blocks sunlight from reaching the moon directly. The only light that reflects off the moon during this time is filtered through the Earth's atmosphere, which scatters blue light and allows red light to pass through. This red light then reflects off the lunar surface, giving it a reddish glow. The more dust or clouds in the atmosphere, the deeper the red hue of the moon will appear. The most recent blood moon occurred on November 8, 2022, and was also known as the Beaver Blood Moon. The next total lunar eclipse, and therefore the next opportunity to witness a blood moon, will be on March 14, 2025. This celestial event will be visible across the Pacific, the Americas, Western Europe, and Western Africa.

- By contrast, a blue moon is a full moon that occurs in a season (typically three months) with four full moons, or twice in a single calendar month. The moon itself doesn't actually turn blue, but it can appear blue when viewed through atmospheric haze containing dust particles (unlike a blood moon, where sunlight is blocked). This phenomenon is rare and unpredictable, hence the saying "once in a blue moon."

- The longest total lunar eclipse on record occurred on June 16, 2000, lasting 106 minutes and 24.6 seconds. The next longest total lunar eclipses are predicted for May 12, 2264, and July 27, 3107, lasting 106 minutes and 13 seconds and 106 minutes and 21 seconds, respectively. The duration of a total lunar eclipse varies depending on the moon's position in its orbit. When the moon is at apogee (its farthest point from Earth), it moves more slowly through Earth's shadow, resulting in a longer eclipse.

- The total lunar eclipse that occurred on February 29, 1504, is known as the Columbus Eclipse, and it famously saved the lives of Columbus and his crew. On Columbus' fourth and final voyage to the New World, his fleet was damaged by shipworms, forcing him to beach his remaining ships on the north coast of Jamaica on June 25, 1503. Initially, the local Arawak people provided food and shelter, but as months passed, tensions rose, and the Arawaks grew reluctant to exchange provisions for what they saw as worthless trinkets. Columbus, facing the risk of starvation, devised a plan. Knowing a total lunar eclipse was imminent on February 29, 1504, he warned the local Arawak chief that his Christian god was angry and would demonstrate his displeasure by turning the full moon "inflamed with wrath." On the night of the eclipse, as the moon reddened and darkened, the terrified Arawaks begged Columbus to restore it. Using an hourglass to time the eclipse, Columbus pretended to consult with his god, and as the moon gradually reappeared, the Arawaks were filled with gratitude and fear and continued to supply Columbus and his crew with food. A rescue ship later arrived in June, and Columbus safely returned to Spain in November 1504.

5.8 Unwelcome Visitors: Asteroid Impacts That Have Caused Disasters

- While nearly 100 tons of meteoric dust enter Earth's atmosphere and burn up every day, smaller asteroids with a diameter of 3 feet (1 meter) hit Earth a few times each year. The asteroid that hit Chelyabinsk, Russia, in 2013 was 62 feet (19 meters) in diameter; events of this scale happen every 60-80 years. Larger impacts, like the 164-foot (50-meter) asteroid that hit Tunguska in 1908, happen every 200-300 years. Asteroids larger than 0.62 miles (1 kilometer) are highly dangerous, with impacts capable of causing catastrophic damage to life on Earth. These occur roughly every 500,000 years. Planet-killing asteroids, larger than 6.2 miles (10 kilometers), like the one that wiped out the dinosaurs 65 million years ago, are extremely rare. Scientists have determined that in the past 600 million years, Earth has been hit by 60 objects with a diameter of 3.1 miles (5 kilometers) or more, averaging one hit every 10 million years.

- The Vredefort Crater is the largest and oldest crater on Earth. It was formed approximately 2 billion years ago when a large asteroid, estimated to be around 6.2 miles (10 kilometers) in diameter, collided with Earth near present-day Johannesburg, South Africa. The impact created the largest known impact crater on the planet that measures 186 miles (300 kilometers) across. This immense crater has been significantly eroded over time, leaving behind the visible "Vredefort Dome" structure. The fact that our planet survived such a catastrophic event is remarkable!

- Approximately 65 million years ago, a large asteroid, estimated to be around 6.2 miles (10 kilometers) in diameter, impacted Earth at a speed over 15.5 miles (25 kilometers) per second near the present-day Yucatán Peninsula in Mexico. The impact released an amount of energy equivalent to 10,000 times the world's current nuclear arsenal, creating a massive crater with a diameter of approximately 124 miles (200 kilometers). Scientists widely believe this impact resulted in sunlight being blocked by dust and debris, which triggered a dramatic global climate change. This led to the mass extinction event at the Cretaceous-Paleogene boundary, wiping out the dinosaurs and over 50% of Earth's species.

- The Popigai crater in Northern Siberia, Russia, was formed about 35 million years ago when a large asteroid, measuring 3-5 miles (5-8 kilometers) wide, impacted Earth. This impact created a massive crater roughly 62 miles (100 kilometers) in diameter. Due to the high carbon content of the soil surrounding the impact site, the intense heat and pressure generated by the collision created a large quantity of "impact diamonds." This abundance of diamonds has earned the crater the nickname "Crater of Diamonds," and you may find tiny diamond granules (0.1-0.4 inches, or 3-10 mm in size) if you visit the site!

- The Sudbury Basin is the world's third-largest impact crater and is located in Ontario, Canada. It was formed over 1.8 billion years ago by a massive comet impact, rather than an asteroid. The comet struck Earth at an estimated speed over 12.4 miles (20 kilometers) per second, creating a crater initially measuring 93 miles (150 kilometers) wide. Over time, erosion reshaped the original crater into an elliptical basin, and its size reduced to approximately 37 by 18 miles (60 by 30 kilometers). Rock fragments from the impact were discovered as far away as present-day Minnesota, hundreds of kilometers from the impact site. The intense heat and pressure generated by the collision melted the rock and concentrated valuable minerals like nickel, copper, and platinum-group metals, leading to the formation of rich mining sites around the crater.

5.9 Galaxies With Radio Signals: Are We Alone In the Universe?

- Are we alone in the universe? Are there other intelligent beings? The short answer is, we don't know. However, scientists are investigating mysterious radio signals from deep space, which some speculate could potentially be a sign of extraterrestrial intelligence.

- The "Wow! signal" is a famous, mysterious radio transmission detected by Ohio State University's Big Ear radio telescope on August 15, 1977, during a search for extraterrestrial intelligence. The signal, lasting 1 minute and 12 seconds, originated near the constellation Sagittarius and was 30 times stronger than the background noise of deep space. It was received at a specific frequency of 1,420 megahertz, the same frequency at which hydrogen, the universe's most abundant element, emits radio waves. This led some astronomers to speculate that it could have been an attempt by extraterrestrial life to communicate with Earth.

- However, since its initial detection, the mysterious "Wow! signal" has never been observed again. In a recent 2024 study, planetary astrobiologist Abel Méndez hypothesized that the signal could have been caused by an intense flare from a magnetar, a type of dense, magnetic star. Méndez suggests that the flare struck an interstellar cloud of hydrogen gas, causing it to emit the radiation detected by the Big Ear radio telescope. Despite this, scientists, including those from NASA, are tirelessly working to collect signals from the universe in the hope of finding signs of intelligent life.

- On the other hand, scientists have detected another type of radio signal from the universe: fast radio bursts (FRBs). FRBs are brief, intense bursts of radio waves originating outside the Milky Way galaxy. These bursts are incredibly energetic, releasing as much energy in a millisecond as the sun does in three days. They can last anywhere from a fraction of a millisecond to several seconds. The first FRB was discovered in 2007, and since then, scientists have detected hundreds of these flashes from distant points across the universe.

- The sources of FRBs remain uncertain. Most scientists believe they are caused by astrophysical phenomena such as black holes or neutron stars. However, some, like Professor Avi Loeb from Harvard, suggest they could be evidence of advanced alien technology. In 2019, a Canadian-led team using the Canadian Hydrogen Intensity Mapping Experiment (CHIME) radio telescope in British Columbia discovered 13 fast radio bursts, including one repeating burst. While most scientists attributed these new bursts to supernova remnants or black holes, Loeb proposed that FRBs could be signals from alien transmitters used to propel spaceships. Scientists hope to gain a more definitive answer as more FRBs are detected.

5.10 Interstellar Migration: The Search for Habitable Planets

- Within our solar system, Mars is considered a potential candidate for future human habitation. It has polar ice caps and evidence of past water channels, suggesting a possible water source. Additionally, a Martian day is roughly the same length as an Earth day, which would facilitate adapting to a Martian circadian rhythm. Although Mars is cold (with a median surface temperature of -85°F, or -65°C) and has a very limited oxygen supply (approximately 1/10,000th of the amount in Earth's atmosphere), scientists believe it could potentially be terraformed to become more suitable for human life.

- Meanwhile, scientists have never stopped searching for habitable planets outside our solar system. Since the first confirmed discovery of an exoplanet (a planet orbiting a star outside the solar system) in 1995, scientists have identified over 5,000 such planets. To find an Earth-like planet habitable to humans, astronomers are looking for small, rocky planets located in the habitable zone of their star, where liquid water can exist. Two of the top candidates that may be habitable are listed below:

- KOI (Kepler Object of Interest) 5725.01, discovered in 2015 through analysis of Kepler space telescope data, is an exoplanet located approximately 3,000 light-years away in the Cygnus constellation. Scientists have determined that KOI-5715.01 harbors conditions even more favorable to life than those found on Earth, leading them to classify the planet as "superhabitable." The planet is nearly twice the size of Earth and likely has a solid surface. It orbits an orange dwarf star, which is more stable than our Sun and has a lifespan of up to 45 billion years, compared to our Sun's 10 billion years. This extended lifespan potentially allows significantly more time for life to develop. With a surface temperature of 11.6 °C, KOI 5715.01 is remarkably close to Earth's average temperature. Scientists are currently studying its atmosphere for molecules like oxygen, carbon dioxide, and methane, which could indicate the presence of life.

- KOI 5554.01, another exoplanet approximately 6.5 billion years old, has a size ranging from 0.72 to 1.29 times the diameter of Earth. It orbits a yellow dwarf star about 700 light-years away, with a year length of 362 days, similar to Earth's. The planet's average surface temperature is about 27°C (80°F), which is suitable for human life. Scientists are currently investigating whether the planet has water and oxygen, key indicators of habitability.

5.11 Everything Is in Motion: We Move Fast Through Space

- The first cosmic velocity refers to the minimum speed an object needs to achieve to orbit a celestial body, like Earth. It's also known as orbital velocity. For Earth, the first cosmic velocity is approximately 4.9 miles (7.9 kilometers) per second. Examples of objects moving at first cosmic velocity include artificial satellites orbiting Earth, the International Space Station, most communication satellites, weather satellites, and any spacecraft maintaining a stable orbit around Earth.

- Second cosmic velocity, also known as escape velocity, is the speed required for an object to completely escape the gravitational pull of a planet and not fall back down. For Earth, the second cosmic velocity is around 7.0 miles (11.2 kilometers) per second. Any spacecraft launched from Earth, like a deep space probe leaving Earth's orbit, is moving at this speed.

- The third cosmic velocity is the initial speed required to escape the combined gravitational fields of Earth and the Sun. The third cosmic velocity is approximately 26 miles (42 kilometers) per second, or 0.014% of the speed of light in a vacuum. A prime example of an object moving at the third cosmic velocity would be a spacecraft that has successfully left the Solar System, like the Voyager probes.

- The Earth's rotation at the equator reaches speeds of approximately 1,037 miles (1,670 kilometers) per hour. This is calculated by dividing the Earth's circumference at the equator by the time it takes to complete one rotation (24 hours). The speed of Earth's rotation decreases as we move further from the equator and closer to the poles. Despite the Earth's rapid rotation, we don't perceive it because the movement is constant, and we move along with it.

- The Moon revolves around the Earth at a speed of approximately 2,288 miles (3,682 kilometers) per hour, or 0.64 miles (1.022 kilometers) per second. It takes 27.3 days for the Moon to complete one orbit around the Earth.

- The Earth revolves around the Sun at approximately 67,000 miles (107,000 kilometers) per hour, or roughly 18.6 miles (30 kilometers) per second. In one year, the Earth travels approximately 584 million miles (940 million kilometers), taking about 365.25 days to complete one orbit.

- Our Milky Way galaxy is like a gigantic spinning pinwheel. Our solar system travels within the galaxy at an average speed of around 483,000 miles (792,000 kilometers) per hour, or roughly 134 miles (220 kilometers) per second, as it orbits the center of the Milky Way galaxy. One full orbit takes approximately 230 million Earth years!

- Scientific measurements indicate that the Milky Way galaxy is hurtling through space at approximately 1.3 million miles (2.1 million kilometers) per hour, which translates to roughly 373 miles (600 kilometers) per second. This speed is measured relative to the cosmic background radiation, a remnant of the Big Bang that serves as a reference point for the universe's expansion. Our galaxy is moving in the general direction of the constellations Leo and Virgo. Despite the Milky Way's high velocity, we don't perceive this motion because we are moving along with it!

- If you're already astonished by the speeds at which we traverse the universe, prepare to be even more amazed. During the Big Bang, the early universe expanded at a mind-boggling pace, far exceeding the speed of light (186,282 miles or 299,792 kilometers per second). Scientists estimate that in the initial stages of the Big Bang, during a phase known as inflation, the universe expanded by a factor of 10^78 in approximately 10^-32 seconds. This translates to an expansion of roughly 62 trillion miles (100 trillion kilometers) in an unimaginably small fraction of a second! It's important to remember that while objects within space cannot surpass the speed of light, the expansion of space itself during inflation was not bound by this limitation.

5.12 Supernova: Fireworks in the Universe

- A supernova is a spectacular event that occurs when a star runs out of fuel and explodes in a brilliant burst of light. Supernovae are among the brightest objects in the night sky and can be billions of times brighter than the sun. They can outshine an entire galaxy for weeks and reach a diameter of several light-years.

- Supernovae are the primary source of heavy elements in the universe. When a supernova occurs, it distributes both stored and newly-created elements throughout space. Scientists have learned a great deal about the universe by studying supernovae, including how stars generate the chemical elements needed for life.

- A supernova occurs somewhere in the universe roughly every 10 seconds. However, in galaxies similar to the Milky Way, only two or three supernovae occur per century. On average, this translates to one supernova every 50 to 100 years in our galaxy; however, this frequency can vary depending on the specific region within the galaxy and the type of supernova.

- A supernova is billions of times brighter than the Sun. A supernova occurring just 32.6 light-years (191.7 trillion miles, or 308.6 trillion kilometers) away would be 15 million times brighter than Sirius, the brightest star in the night sky.

- Due to their extreme brightness, supernovae can be observed from vast distances. According to current astronomical data, the most distant supernova ever detected is DES16C2nm, located approximately 10.5 billion light-years from Earth. This superluminous supernova (SLSN) was first spotted by the Dark Energy Survey in August 2016.

- For a star with a mass at least 20 times that of the sun, a supernova explosion can leave behind a remnant with sufficient mass to form a black hole. In these massive stars, the gravitational forces overwhelm the neutron degeneracy pressure, causing the star's core to collapse into a black hole. If the stellar remnant remains below this critical mass threshold, it will instead form a neutron star. The heaviest known neutron star is approximately 2.5 times the mass of the sun, whereas the lightest known black hole is around 5 solar masses.

- We know black holes are strange, but neutron stars also have unusual characteristics. They contain the densest material that we can directly observe in the universe, with densities reaching around 10^17 kilograms per cubic meter. Just one teaspoon of neutron star matter would weigh as much as Mount Everest, which weighs billions of tons! They also spin incredibly fast. The fastest known neutron star, PSR J1748-2446ad, can spin around 43,000 times per minute (716 times per second), which is nearly as fast as a blender blade. Neutron stars also boast the strongest known magnetic fields in the universe. Their magnetic fields, measuring 10^12 to 10^15 Gauss, are trillions to quadrillions of times stronger than Earth's magnetic field (0.5 Gauss), and a type of neutron star called a "magnetar" can have magnetic fields even a thousand times stronger than that! Such powerful magnetic fields can significantly affect nearby space.

- The brightest supernova ever recorded, called "SN2016aps," was discovered in 2016 by the Panoramic Survey Telescope and Rapid Response System (Pan-STARRS) in Hawaii. The peak luminosity of SN2016aps (absolute magnitude −22.35) was 74 billion times brighter than the Sun (absolute magnitude +4.83), and it was 500 times brighter than typical supernova explosions observed to date! Its energy output is a staggering 1.1 quintillion (10^18) times that of the Sun, making it one of the most energetic events ever observed. The supernova occurred in a galaxy roughly 3.6 billion light-years away and was so luminous that it outshone the light of its surrounding galaxy.

5.13 The Origin and Fate of the Universe: Where Will It All End?

The Origin of the Universe

- The prevailing scientific view is that the universe originated with a "Big Bang" approximately 13.8 billion years ago, expanding from an extremely dense, single point. This expansion continues today at an accelerating rate.

- Immediately following the Big Bang, the universe existed as a superheated "soup" of particles, with temperatures estimated to have reached around 10^32 Kelvin. Within the first second of existence, the universe rapidly cooled to temperatures around 10 billion degrees Celsius (still very hot!). By comparison, the temperature at the Sun's surface is only about 10,000 degrees Fahrenheit (5,600 degrees Celsius), while its core reaches approximately 27 million degrees Fahrenheit (15 million degrees Celsius).

- The extreme temperatures of the early universe facilitated the continuous creation and annihilation of particles like electrons and positrons. During the initial few minutes, the universe generated most of the hydrogen and helium that exists today, along with trace amounts of lithium. Hydrogen and helium are the most abundant elements, making up approximately 98% of the visible universe.

- Within a tiny fraction of a second (10^-32 seconds) after the Big Bang, the universe underwent a period of rapid expansion known as cosmic inflation. During inflation, the universe's linear dimensions expanded by a factor of at least 10^26, resulting in a volume increase of 10^78. To put this in perspective, it's like expanding something as small as half the width of a DNA molecule to roughly 10.6 light-years across (62.3 trillion miles, or 100 trillion kilometers)!

- The period beginning about 380,000 years after the Big Bang and lasting until the formation of the first stars and galaxies, which was estimated to be around 1 billion years after the Big Bang, is known as the cosmic dark ages. During this era, the universe was filled with neutral hydrogen gas that obscured light sources. This made the universe appear dark.

- The observable universe is about 93 billion light-years in diameter, but the true size of the universe is unknown. This is because we can't see past the edge of the observable universe. Within the observable universe, scientists estimate that there are between 100 billion and a trillion galaxies, and 200 billion trillion stars. Compared to the universe, our Earth is like a grain of sand lost in a sprawling desert!

The End of the Universe

- The potential end of the universe could take the form of a "Big Freeze," where it continues expanding indefinitely, or a "Big Crunch," where gravity eventually pulls everything back together. The outcome depends on the universe's density and the amount of dark energy present.

- The Big Freeze, also known as the Big Chill or heat death, is a widely accepted theory. It proposes that the universe will expand forever, dispersing energy until all regions reach a uniform temperature. This would cause particle interactions to stop, leaving the universe in a frozen state.

- This process is likely to take over a trillion years. Initially, star and galaxy formation would stop, and existing stars would gradually run out of fuel and stop shining. Black holes would then consume any remaining matter, and the universe would enter a Dark Era. Ultimately, the universe would reach a state of maximum entropy, with everything evenly distributed and no energy gradients to drive any further events. Imagine time becoming an endless void, where virtually nothing happens due to the lack of energy in the universe!

- In a different scenario, if there is enough gravity, the universe could eventually collapse back in on itself, similar to a collapsing star forming a black hole. This would ultimately cause the universe to shrink to an infinitely small point known as a singularity, an event potentially followed by another Big Bang and the reformation of a new universe.

- In both theories, the mysterious dark energy is believed to be the driving force behind the accelerating expansion of the universe. However, if dark energy causes the expansion of the universe to accelerate too rapidly, it could ultimately rip the universe apart, reducing it to its fundamental components like atoms. Which way do you think our universe will end?

Chapter 6

The Mysterious Brain

6.1 Brains: The Most Powerful Computers We Ever Know

- The human brain is the most complex known structure in the universe, containing roughly 86 billion neurons. Each neuron can form thousands of synaptic connections with other neurons, creating a vast network of trillions of interconnections. This intricate network far surpasses the complexity of even the most advanced supercomputers.

- The human brain possesses remarkable processing capabilities, handling vast amounts of information simultaneously. While estimates vary, recent research suggests its computational power may reach the exaflop range (one quintillion, or 10^18 calculations per second), rivaling or even exceeding the raw processing power of some supercomputers.

- Despite having the computational power of a supercomputer, the human brain operates on only about 20 watts of energy—roughly equivalent to the power needed to run a dim light bulb. In contrast, one of the world's most powerful supercomputers, the Oak Ridge Frontier, can achieve similar exaflop computing speeds (10^18 calculations per second) but requires 20 megawatts to do so. This is a million times more power than the brain uses.

- The human brain can store about 2.5 petabytes of information, which is equivalent to 2.5 million gigabytes of digital memory, or more than 2,500 1TB solid-state hard drives (SSDs). This is roughly comparable to the amount of data stored on the World Wide Web. The brain's storage power is derived from the vast number of connections within it, which amounts to 100 trillion connections between its

100 billion neurons.

- Believe it or not, the vast majority of neurons in our brains develop before we're born. The peak of neuron production occurs during the middle stages of gestation (between the 5th and 20th weeks), during which our brains generate an astounding 50,000 to 100,000 new nerve cells per second. Once formed, neurons migrate to their designated locations in the brain and begin to differentiate into specialized types. They then start forming connections with each other called synapses, which are crucial for neural communication.

- Despite the continual structural refinement of our brains after birth, we experience a gradual loss of neurons. Scientists estimate that the total number of neocortical neurons decreases by 9.5% between the ages of 18 and 93[1]. This translates to an average loss of approximately 85,000 neurons per day, or roughly one neuron per second. However, some brain regions, like the hippocampus (responsible for memory formation and consolidation) and the olfactory bulb (involved in odor detection), continuously generate new neurons. Researchers estimate that the hippocampus produces around 700 to 1,500 new neurons each day throughout adulthood[2], and exercise likely promotes the generation of new neurons.

6.2 Male and Female Brains: Do Brains Have a Sex?

- The male and female brains indeed show differences in both structure and function. Research has found that male brains are, on average, about 10% larger than female brains. Stanford scientist Nirao Shah suggests that neural circuits governing innate behaviors, such as mating and parenting, are likely hard-wired and show sex-specific variations.

- Scientists find that women generally demonstrate superior performance in various measures of verbal ability, including reading comprehension, writing skills, and verbal fluency. The exception is in verbal analogies, where men tend to perform better. Additionally, women excel in tests of fine-motor coordination and perceptual speed. In contrast, men have superior visuospatial abilities. They are better at visualizing how complex shapes rotate in space, determining angles from the horizontal, tracking moving objects, and aiming projectiles.

- Women are typically more adept at retrieving information from long-term memory, and they retain stronger, more vivid emotional memories. They tend to have richer, more detailed memories of emotional events and recall these experiences faster than men. This superior memory for emotional events may be related to differences in the functioning of the amygdala. Additionally, females tend to rely more on landmarks for navigation, whereas males typically use "dead reckoning," estimating their position based on distance and direction traveled.

- Interestingly, research indicates that boys and girls display distinct toy preferences from a young age. For instance, boys often prefer toys with wheels, whereas girls tend to favor plush toys. This suggests that these preferences may be inherent rather than solely shaped by cultural influences. Notably, even differences in spatial-visualization ability can be observed in infants as young as 2 to 3 months old.

- The prevalence of cognitive and neuropsychiatric disorders also differs between the sexes. For example, women are twice as likely to experience clinical depression or post-traumatic stress disorder (PTSD), while men are more likely to develop alcoholism, drug dependence, or schizophrenia. Boys have a roughly tenfold higher rate of dyslexia and are four to five times more likely to be diagnosed with autism spectrum disorder[3].

6.3 Extraordinary Brain Capacities: More Powerful Than We Realize

Synesthesia

- The saying "I know how you feel" can be taken literally by hyper-empathetic individuals with a rare condition called mirror-touch synesthesia. These individuals actually feel the sensation of being touched when they witness another person being touched. In 2007, scientists reported a woman with mirror-touch synesthesia who felt a touch on her right cheek when she saw someone standing in front of her, face to face, being touched on the left cheek[4]. Interestingly, if she stood next to the person being touched, the sensation would be on the same side as the person being touched.

- The mirror-touch synesthesia can go to an extreme and become mirror-pain synesthesia. Dr. Joel Salinas, who experienced music and numbers as colors as a kid, can feel others' pain. As a medical student, Joel felt the incision of the operating knife in his own abdomen when he watched a teenage boy on an operating table. In 2008, Joel felt chest compressions on his own chest and the breathing tube scraping his throat when he witnessed an emergency rescue of a cardiac arrest patient. When the patient died, an eerie silence and absence of sensations enveloped him[5]. However, mirror-touch synesthesia also helped Joel notice patients' needs quickly and better care for his patients.

- This condition is related to the activity of mirror neurons, brain cells that fire electric signals not only when performing an action but also when observing the same action performed by another. Scientists believe we're all born with synesthesia, which diminishes as unnecessary brain connections are pruned. For individuals with mirror-touch synesthesia, which affects about 1.6% of people, these mirror neurons seem to be in overdrive. This supports the idea that empathy develops through mentally simulating the experiences of others.

Savant Syndrome

- Savant syndrome is a rare condition where individuals with developmental disabilities, like autism, or brain injuries, exhibit extraordinary skills in specific areas. These skills, often termed "savant abilities," typically manifest in domains such as art, memory, arithmetic, music, or spatial abilities. The most common type of savant is thc calendrical savant, often referred to as a "human calendar," who can accurately calculate the day of the week for any given date. Enhanced memory is considered a core "superpower" underlying these savant abilities. While savant syndrome is more prevalent in males, with a 6:1 male-to-female ratio, it affects only an estimated one in a million people. The condition is not classified as a mental disorder but is thought to arise from the brain's ability to reorganize and compensate for impairments.

- Daniel Tammet is an English writer and savant born on January 31, 1979 and is known for his extraordinary memory capacity. Tammet perceives each positive integer up to 10,000 with unique shapes, colors, textures, and feelings, and he describes numbers with vivid imagery. He finds 289 ugly, 333 attractive, and pi beautiful. He holds the European record for reciting pi to 22,514 digits from memory in five hours. Additionally, Tammet is a polyglot and speaks eleven languages: English, Estonian, Finnish, French, German, Lithuanian, Esperanto, Spanish, Romanian, Icelandic, and Welsh. He learned conversational Icelandic in just one week, which was astonishing considering the difficulty of the language.

- Laurence Kim Peek was an American savant born on November 11, 1951, in

Salt Lake City, Utah. Kim was known as a "megasavant" for his extraordinary memory. As a kid, Kim faced challenges with motor skills. He couldn't walk until he was four and struggled with tasks like buttoning his shirt. However, despite social difficulties and congenital brain abnormalities, Kim became famous as the inspiration for the character Raymond Babbitt in the 1988 movie "Rain Man." His memory capacity was remarkable; he could memorize an entire book in about an hour and accurately recall the contents of at least 12,000 books. Kim read books rapidly, scanning left pages with his left eye and right pages with his right eye. His exceptional abilities were attributed to unique neural connections due to the absence of the corpus callosum, a brain structure connecting the left and right cerebral hemispheres.

- Stephen Wiltshire, born on April 24, 1974, is a British architectural artist and autistic savant known for his extraordinary ability to draw detailed cityscapes from memory after seeing them just once. Diagnosed with autism at age three, Stephen was nonverbal in his early years. His fascination with sketching landmarks began around the age of seven. Stephen's talent has led him to create detailed panoramic drawings of cities including London, New York, Tokyo, Rome, Hong Kong, Frankfurt, Madrid, Dubai, and Jerusalem. One notable work is his 18-foot-long (5.5 m) drawing of New York City completed after a brief 20-minute helicopter ride, which spans Manhattan, the Hudson shoreline of New Jersey, the Financial District, Ellis Island, the Statue of Liberty, and Brooklyn. He also produced a 32.8-foot-long (10 m) panoramic memory drawing of Tokyo in just seven days. In 2006, Stephen was made a Member of the Order of the British Empire (MBE) for services to art and opened his permanent gallery in the Royal Opera Arcade in London.

- Jason D. Padgett, born in 1970 in Anchorage, Alaska, is an American artist diagnosed with acquired savant syndrome. As a young man, he dropped out of Tacoma Community College and worked as a salesman for his father's futon company. On September 13, 2002, after a brutal attack outside a karaoke bar in Tacoma, Washington, Jason sustained a severe concussion. Following this traumatic event, he began to perceive the world through a mathematical lens. Water appeared as tangent lines, and light as rays made up of spirals and lines. This newfound perspective led him to create intricate geometric drawings, which helped him manage his obsessive-compulsive disorder (OCD) and post-traumatic stress disorder (PTSD). His extraordinary ability caught the attention of Dr. Darold Treffert, a renowned expert on savantism, who diagnosed him with acquired savant syndrome. Jason is one of only 40 people in the world with this condition, in which remarkable talents in math, art, or music emerge following a brain injury or disease.

- Derek Amato, born on November 19, 1966, was a corporate trainer in Colorado. His life changed dramatically on October 27, 2006, when he dove into a swimming pool at a friend's party and struck his head on the bottom, sustaining a severe concussion and losing some hearing in one ear. However, a few days after the accident, music flooded into his head. He sat down at a piano and began to play masterpieces, even though he had barely been able to play the instrument before his injury. He won the 2007 Independent Artist of the Year award by the LA Association of Independent Artists and was called "Rainman Beethoven" for his acquired savant syndrome.

- Orlando L. Serrell, born in 1968, became an "acquired savant" after a life-changing accident on August 17, 1979. At the age of ten, Orlando was struck by a baseball on the left side of his head and lost consciousness. When he regained consciousness, he did not seek medical treatment. Orlando suffered from a persistent headache; however, upon recovering, he began to notice an extraordinary ability to perform complex calendrical calculations. He could determine the day of the week for any given date since the accident, and he could recall the weather and specific details of each day. Over time, his memory skills evolved into hyperthymestic memory, enabling him to remember even the minutest details of daily activities with remarkable clarity. Scientists who have studied Orlando's extraordinary abilities have found that he engages in unconscious calculation without memorizing the calendar. Orlando's story is a testament to the incredible potential of the human brain and how it can adapt in unforeseen ways following trauma.

- Scientists found the phenomenon of acquired savant syndrome involves specific brain changes. After an injury, the brain undergoes a process known as the "three Rs": recruitment, rewiring, and release. Initially, there is a surge of electrical activity in the remaining healthy brain tissue, often in the right hemisphere. The brain then forms new neural connections, linking previously unconnected regions. This rewiring process leads to the release of dormant capacities, allowing individuals like Jason to access extraordinary skills. In support of this theory, scientists at the University of Sydney were able to temporarily induce savant-like abilities in participants by decreasing activity in the left hemisphere while increasing it in the right.

- The acquired savant syndrome underscores the idea that our brains are not a blank slate. Rather, they may come pre-equipped with inherent knowledge and abilities. Savants can access these innate predispositions, revealing the brain's extraordinary potential for creativity and understanding in music, art, and mathematics.

Human Calculator

- Shakuntala Devi (November 4, 1929 – April 21, 2013) was an Indian mental calculator, astrologer, and writer known for her extraordinary talent in mental calculation. Despite having no formal education, she showed remarkable arithmetic abilities and was known as the "Human Computer." Shakuntala demonstrated her mental arithmetic talents in several countries. She determined the 23rd root of a 201-digit number in just 50 seconds in 1977 at Southern Methodist University, outperforming the computer program at the US Bureau of Standards. She also correctly calculated the cube root of 61,629,875 and the seventh root of 170,859,375 in seconds at the University of California, Berkeley, in 1988. On 18 June 1980, she demonstrated her exceptional skills by multiplying two randomly selected 13-digit numbers—7,686,369,774,870 × 2,465,099,745,779—chosen by the Department of Computing at Imperial College London. She correctly provided the answer, 18,947,668,177,995,426,462,773,730, in only 28 seconds, which was recorded in the 1982 Guinness Book of World Records.

- Aaryan Shukla, born in Maharashtra, India, is a 13-year-old prodigy renowned for his extraordinary mental calculation abilities. His rapid mental arithmetic earned him a Guinness World Record in 2024. On the set of the Italian TV series *Lo Show Dei Record*, Aaryan stunned the audience by mentally adding 50 five-digit numbers in just 25.19 seconds—an average of one addition every 0.5 seconds. As the numbers flashed before him on a large screen too quickly for the audience to even process, Aaryan mentally calculated the total and delivered the correct answer. Aaryan also won the prestigious Mental Calculation World Cup championship in 2022, held in Paderborn, Germany. In 2021, at the age of 10, Aaryan made headlines by winning gold medals in both the open and junior categories at the Mind Sports Olympiad (UK), an intense 3-hour online competition. Competing against seasoned mathematicians, Aaryan tackled a series of challenging problems, including adding 10-digit numbers, multiplying 7-digit numbers, and extracting square and cube roots. Without pen, paper, or calculator, Aaryan consistently outpaced his competitors with his astonishing mental power. He is the youngest known human calculator.

6.4 Mysterious Brain Abnormalities

Fearless Warrior

- S.M., known as the "patient without fear," is a remarkable case in medical history due to her inability to experience fear. In a 2011 study, researchers at the University of Iowa tested S.M.'s extraordinary condition through a series of scenarios designed to elicit fear[6]. These tests included exposure to live snakes and spiders, a tour of one of the world's most haunted houses, and viewing clips from terrifying films. Despite her expressed aversion to snakes, S.M. held and interacted with a snake for over three minutes without fear. She also showed no fear in the haunted house or when watching the horror films; instead, she found them exciting and entertaining. In her real life, S.M. has faced life-threatening situations, including being held at knifepoint by a strong man in a dark park, but she never felt fear. Scientists found that S.M. has bilateral lesions in a brain structure called amygdala as a result of a rare inheritable condition known as Urbach-Wiethe disease. This condition, characterized by calcium deposits in the amygdala, has led to a complete inhibition of her fear response.

- Surprisingly, amygdala impairment does not completely eliminate all fear responses. In a 2013 study, scientists found that S.M. reported experiencing fear for the first time since childhood during a 35% CO2 inhalation challenge[7]. Inhalation of a high concentration of CO2 leads to suffocation, an experience that also triggered a panic attack in S.M. This was the first documented instance of S.M. feeling fear in any setting, laboratory or otherwise. The researchers conducted the same test on two additional patients who also had amygdala lesions due to Urbach-Wiethe disease, and both patients similarly reported experiencing fear during the CO2 challenge, confirming the findings.

- The fearless Jordy. Jordy Cernik, a British broadcast producer, suffered from a rare condition called Cushing's syndrome, which causes excessive cortisol production in the body. Since his initial diagnosis in 2005, he underwent several brain surgeries to reduce cortisol levels. The final surgery in 2010 completely removed his adrenal glands and, unexpectedly, he lost the ability to feel fear. Despite being a fearful person before the surgeries, Jordy no longer felt fear or any associated emotions like excitement or nervousness. Whether rappelling down a 127-meter tower or sitting on a tightly packed plane at 19,000 feet, his heart rate remained unchanged. While many view this fearlessness as a superpower, Jordy struggles with the lack of anxiety for his children's safety and misses the excitement and joy of daily activities like booking a family holiday. However, Jordy remains connected to his family through love and cherishes the memories he had before the surgeries.

Come On, Your Smile Is Fake

- We smile a lot, and some smiles are genuine while others are not. However, few people know that genuine and fake smiles are controlled by different neural pathways in our brains. A genuine smile is controlled by cells in the basal ganglia, which generate involuntary facial expressions. In contrast, a forced or fake smile is managed by the motor cortex in the front of the brain, which produces skilled voluntary movements. In 1862, French anatomist G. B. Duchenne observed that genuine smiles involve the orbicularis oculi muscle around the eyes. They often also show raised cheeks, crow's feet wrinkles, and a slight lowering of the eyebrows. A genuine smile is now known as the Duchenne smile.

- Evidence supporting the existence of two distinct smile circuits comes from studying brain-damaged patients. For instance, when a stroke affects the right motor cortex (which governs movement on the left side of the body), a patient may only be able to produce a partial smile on the right side of their face when asked to do so. However, a spontaneous, joyful emotion, such as seeing a loved one, will trigger a full, symmetrical smile because the basal ganglia, responsible for genuine smiles, remain unaffected. Conversely, a stroke impacting the basal ganglia on one side may allow a patient to produce a forced, symmetrical grin on command, but their natural smile will appear as a half smile, highlighting the damage to the emotional, genuine smile circuit[8]. These cases illustrate that our brains possess two separate circuits for controlling voluntary and emotional smiles.

Split Brain: Split Self?

- Our brains are divided into two hemispheres, left and right, each controlling the opposite side of the body. The left hemisphere is primarily responsible for language and rhythm, while the right hemisphere is more involved in non-verbal tasks such as recognizing objects, emotions, and melody. The two hemispheres are connected with a thick bundle of nerve fibers called the corpus callosum. In some cases of severe epilepsy, the corpus callosum is surgically severed to treat seizures. This procedure, however, prevents the brain's hemispheres from communicating properly, which can lead to unusual split-brain symptoms.

- A middle-aged woman sought medical attention because her left hand would occasionally act on its own and try to strangle her. She frequently had to use her right hand to restrain her left hand or even sit on it to stop it. Despite this alarming behavior, she appeared otherwise normal, showing no signs of neurological problems. Initially, she was misdiagnosed with mental illness or hysteria. However, one doctor found that her symptoms were caused by damage to the corpus callosum. This damage, likely due to a stroke, disrupted the normal inhibitory signals from the logical left hemisphere to the more emotional right hemisphere. As a result, the right hemisphere, which controlled her left hand, was no longer kept in check and acted independently, showing what appeared to be suicidal tendencies. This caused the woman's left hand to attempt to strangle her in a seemingly uncontrollable manner[8].

- Researchers also discovered that when an object is shown to a split-brain patient's right eye (which sends signals to the left hemisphere), the patient can easily name it, as the left hemisphere is dominant for language. However, when shown to the left eye (which sends signals to the right hemisphere), the patient cannot verbally describe the object but can still select the corresponding object with their left hand. In a famous experiment with a split-brain patient named "Joe", the word "Texas" was flashed to his left eye (processed by the right hemisphere). His left hand (also controlled by the right hemisphere) then drew a cowboy hat. However, when Joe was asked why his left hand drew the hat, he could not explain because his language center (located in the left hemisphere) had not received any information about the word[9].

- In these split-brain patient cases, it appears as though there are two separate conscious minds within the same body, one governing speech and the other non-verbal actions like pointing or drawing. Such studies illuminate how different brain regions can function independently when they can't communicate with each other, leading to some fascinating behaviors and insights into how our consciousness might be divided across the brain.

Blindsight: Unconscious Observing

- We rely heavily on our vision to perceive the world, and around 90% of the information transmitted to our brain comes from visual input. When we observe something, light enters our eyes, and images are focused onto the retina, which contains photoreceptor cells. These cells convert light into electrical and chemical signals that are transmitted through the optic nerve to the primary visual cortex, which is located in the occipital lobe at the back of the brain. The brain then processes these signals, creating a visual image. Any disruption in this visual pathway, including damage to the retina or visual cortex, can result in blindness.

- However, doctors encountered blind patients who showed unusual behaviors. One well-studied patient, known as DB, sustained damage to his visual cortex during brain surgery to treat headaches. Although his eyes were healthy, DB was unable to consciously see anything in the left half of his visual field. Everything to the left of his nose was completely invisible, as if a curtain had been drawn over that side of his vision. Despite his belief that he couldn't see anything in this area, his actions suggested otherwise. He could reach out and grab an object in his blind spot, as if guided by a form of "second sight," even though he had no conscious awareness of seeing it. When psychologists placed objects in front of DB's blind spot and asked him to point to them or determine their orientation, he insisted he couldn't see anything. Yet, when asked to guess, he was surprisingly accurate most of the time — far exceeding random chance[10]. Although DB's conscious vision was impaired, his brain was still processing some visual information subconsciously, enabling him to perform certain tasks without being aware of seeing.

- Another patient, known as TN, completely lost his vision after suffering two strokes that damaged the visual cortex in both hemispheres of his brain. This resulted in total blindness. Despite this, an extraordinary experiment conducted by Tamietto and Weiskrantz's team in 2008 revealed that TN still possessed a remarkable, subconscious ability to navigate his surroundings. In the experiment, the researchers removed TN's cane and placed various pieces of furniture in a long corridor, creating obstacles that could potentially trip him. TN was asked to walk down the hallway, and even though he reported being unable to see, he skillfully navigated around the objects on his first attempt, avoiding all the obstacles without bumping into them. To the researchers' surprise, TN claimed afterward that he was completely unaware of having seen or avoided the objects. He believed he had simply walked straight through the hallway, with no conscious awareness of his movements[11]. This case demonstrates that although TN had lost all conscious vision due to damage to his visual cortex, other brain pathways outside the visual cortex—called extra-striate pathways—were still functioning and guiding his behavior.

- Research on blindsight patients emphasizes that awareness doesn't tell the whole story of how our brains work. We often think we make conscious decisions, but our brains may decide before we're aware. These patients show that even without conscious sight, the brain can still process visual information and guide their actions. This suggests our subconscious mind can handle complex tasks, and consciousness may simply summarize the brain's processed information. These findings are leading scientists to reconsider how we understand consciousness, and studying patients like TN could eventually help unlock the mysteries of its true nature.

Phantom Limbs: They Are There

- A phantom limb is the experience of still feeling a limb that has been amputated. It can include feeling pain, pressure, itching, or other sensations in the missing limb. Phantom limb sensations can feel very real, and it can be easy to think the limb is still there. Patients may feel the ghostly presence of fingers and toes, and can even perform actions like reaching out and grabbing objects. Phantom limb sensations are a common experience, affecting about 90% of amputees.

- In a striking case of phantom limb sensations, a patient who had undergone an arm amputation began experiencing a phantom arm shortly after the surgery, a common occurrence. However, after a few weeks, he started feeling a strange, gnawing sensation in his phantom limb. Curious about the cause, he asked his doctor what he had done with his amputated arm, but his doctor was unaware and suggested he contact the surgeon. After a few rounds of phone calls, the patient discovered that the amputated arm had been sent to the pathology lab instead of being incinerated. Unfortunately, because the pathology lab already had so many amputated arms, they had buried the patient's arm in the garden behind the hospital. When the patient exhumed the arm, he found it infested with maggots! The arm was subsequently incinerated, and the patient's phantom gnawing sensation disappeared[8]. Does this indicate a lingering connection between the patient and his amputated arm? The answer remains elusive.

- Besides phantom limbs experienced by amputees, in rare cases, patients with normal limbs can experience the presence of an extra limb, known as a supernumerary phantom limb (SPL). In one case study, a 78-year-old male patient experienced an SPL after a right basal ganglia hemorrhage (bleeding). Following a sudden headache and weakness on the left side of his body, a CT scan revealed a brain hemorrhage extending into the thalamus. On the second day of hospitalization, the patient began feeling a phantom arm beside his paralyzed left arm. He could not see or touch this arm but felt it vividly, reporting that it resembled his real arm in size and shape. The phantom arm sensation was constant and unaffected by closing his eyes or looking at his actual arm. Interestingly, the patient could move the phantom arm voluntarily, including waving, gripping, and touching objects, with strength comparable to his healthy right arm. However, he could not extend the phantom arm far or feel gloves placed on his real left hand when attempting to feel them with the phantom. The patient did not experience any pain or discomfort in either his real, paretic arm or the phantom arm. His phantom arm disappeared after 3 weeks[12].

- Scientists later discovered that phantom limb sensations emerge in amputees through a process called cortical remapping in the brain. After an amputation, the brain no longer receives sensory input from the missing limb. However, the brain areas that previously controlled the limb don't remain inactive. Instead, these regions are "taken over" by nearby areas on the brain's sensory map. In the case of a patient named Tom, sensations from his missing hand were triggered when certain parts of his body, like his face or upper arm, were touched[8]. This occurred because the sensory areas for the face and upper arm are located next to the hand area in the brain's sensory cortex. After the amputation, the sensory nerves that typically activated the face or upper arm began to "invade" the vacant hand area of the brain. This remapping process caused touch sensations from the face or upper arm to be "felt" in the missing hand, creating a phantom limb sensation.

- Interestingly, doctors have discovered that mirrors can sometimes alleviate disturbing phantom limb pain, particularly after a hand or arm amputation. By positioning a mirror to reflect the patient's healthy limb, and having the patient move that limb, the illusion is created that the phantom limb is also moving. This technique tricks the brain into perceiving the missing limb as present, thus providing positive feedback to the motor cortex and helping reduce pain. Recently, other approaches like cognitive multisensory rehabilitation (CMR) have been developed to help restore multisensory body representation and reduce phantom limb pain in patients[13].

The Man Without Memory

- H.M. (1926–2008) is one of the most famous cases in neuroscience for memory research. At the age of 7, H.M. suffered a bicycle accident that resulted in severe epilepsy. By the time he was 27, his seizures had become so debilitating that he could no longer lead a normal life. In 1953, he underwent an experimental brain surgery to control his severe seizures. The surgery removed significant portions of his temporal lobe, including both sides of a brain structure called the hippocampus. While the surgery was successful in reducing his seizures, doctors discovered that H.M. had lost the ability to form new long-term memories, a condition known as anterograde amnesia. H.M. could only remember events prior to his brain surgery, while his short-term memory (like remembering a phone number to make a call) and procedural memory (which is involved in learning skills like riding a bike) remained intact. Remarkably, H.M. could still solve crossword puzzles using knowledge he had acquired before the surgery in 1953. The case of H.M. provided a foundational understanding of the role of the hippocampus in memory formation.

- In contrast, other patient cases demonstrate that memory loss can also be retrograde, meaning the loss of older memories. One notable case is that of S.B., a former NFL player who sustained a head injury in 2008 after slipping in a bathroom. The injury resulted in the near-total loss of his memories prior to the incident, a condition known as retrograde amnesia. He was unable to recall 46 years of his life, including his identity, his wife, his children, and his career. He even lost comprehension of the word "wife" and had no memory of significant life events such as his football career or the births of his children. He could not recognize his family or home and even forgot how to start his car. Over time, S.B. has been able to piece together his past through stories and photos shared by his wife and friends, but he remains unable to fully recover his memories. Doctors determined that his brain had no blood flow to the right temporal lobe (where the hippocampus is located), a key area for memory storage, making his condition likely irreversible[14]. Unlike anterograde amnesia, where new memories cannot be formed, S.B. can create new memories but cannot recall most of his past life.

- Thanks to these patients, scientists have been able to discover the brain regions critical for memory formation and retrieval. We now know that damage to the brain's medial temporal lobe, particularly the hippocampus, is primarily responsible for both anterograde and retrograde amnesia, as this region is critical for memory consolidation and retrieval. In anterograde amnesia, where patients cannot form new memories after the brain damage occurs, the impairment is most strongly linked to the hippocampus itself. In retrograde amnesia, however, where patients lose memories from before the brain injury, the damage can be found in surrounding areas in the medial temporal lobe, such as the parahippocampal gyrus.

- In fact, memory loss is more common than we realize. It is prevalent in older adults and patients with dementia (like those with Alzheimer's Disease). An estimated 40% of people over age 65 and nearly all Alzheimer's patients experience some form of memory loss. However, scientists propose that in these cases, memories are not truly lost; rather, we lose the ability to retrieve them. If this is the case, is there a way to regain access to these memories?

- Recent studies suggest this is possible. In 2015, MIT neuroscientists discovered that memories "lost" to amnesia can be reactivated using light, through a technique called optogenetics. The MIT team, led by Susumu Tonegawa, used optogenetics to manipulate neurons with light. They demonstrated that memories are stored in groups of brain cells called memory engram cells, which undergo physical or chemical changes when a memory is formed. The researchers showed that although these memory cells are still present in amnesiac subjects, the subjects cannot naturally recall the memories without the necessary synaptic strengthening, known as long-term potentiation (LTP). The team experimented with mice, blocking LTP with a drug called anisomycin to prevent memory consolidation after the mice formed new memories. The mice were unable to recall these memories naturally; however, when the scientists activated the engram cells directly with light, the mice fully recalled the memory, proving that it wasn't lost but just inaccessible[15]. This groundbreaking research addresses a long-standing debate in neuroscience about amnesia, and shows it is possible to treat memory loss conditions if the right triggers are available.

Frontal Lobe And Personality: No Longer Himself

- Phineas Gage (1823–1860), an American railroad worker, experienced a traumatic brain injury in 1848 at the age of 25. The injury occurred while Gage was preparing explosives for railroad construction. A spark ignited the powder, propelling an iron rod through his left cheek, behind his left eye, and out through the top of his skull, causing significant damage to his left frontal lobe. Although Gage survived the injury and retained his memory and general intelligence, his personality and behavior underwent a dramatic transformation. Before the accident, he was known for being responsible, hardworking, and well-liked. However, afterward, he became impulsive, irreverent, and prone to inappropriate behavior, leading friends to remark that he was "no longer Gage." Despite the severity of his injury, Gage lived for another 12 years, even working as a stagecoach driver in Chile. His skull and the iron rod are now on display at Harvard University. Gage's case remains one of the most renowned in neuroscience and psychology, as his injury provided crucial insights into the role of the frontal lobe in decision-making, emotional regulation, and social behavior.

- Our understanding of the frontal lobe's role in personality also comes from the practice of lobotomy. A lobotomy is a surgical procedure that severs the connections between the frontal lobe and the thalamus. It was once used to treat mental illnesses that were unresponsive to standard treatments, such as obsessive-compulsive disorder (OCD), severe depressive illness, psychosis, schizophrenia, manic depressive psychosis (now called bipolar disorder), chronic neurosis, and psychopathic personality. The idea was that by disconnecting the "overactive" parts of the brain, doctors could reduce symptoms like anxiety, mood swings, and aggression. Pioneered in the 1930s and 1940s, lobotomy was commonly practiced in the mid-20th century. However, it soon drew criticism for causing severe and often irreversible side effects, including profound changes to patients' personalities and intellectual functioning. Following the introduction of the first antipsychotic medication, chlorpromazine (initially developed as a "tranquilizer"), in 1951, the number of lobotomies performed decreased dramatically before the procedure was largely abandoned.

- The case of Howard Dully exemplifies the potential negative impacts of lobotomy. Howard Dully underwent a transorbital lobotomy at the age of 12, in 1960. Dr. Freeman performed the operation, and it was quite controversial because Dully was not diagnosed with a severe mental illness; his stepmother described him as difficult to manage. She claimed he was defiant, and Freeman agreed to the procedure, even though Dully didn't meet typical criteria for such drastic surgery. The lobotomy targeted Dully's prefrontal cortex, the brain region responsible for decision-making, emotional regulation, and impulse control. By severing connections in this area, the surgery aimed to "calm" him, but it led to lasting repercussions. After the surgery, Dully experienced emotional numbness and a lack of engagement with the world. He suffered from a decline in cognitive function, including difficulties with focus and critical thinking. Additionally, he struggled to form relationships, and his social behavior became unpredictable[16]. Dully faced challenges with alcoholism, homelessness, and depression. However, over time, he showed remarkable resilience and gradually achieved greater stability in his life. Dully later documented his experiences in his memoir, "*My Lobotomy*," with the goal of educating others about the risks of psychosurgery and advocating for improved mental health care.

- Our current knowledge of the frontal lobe helps us better understand adolescent emotional development and behavior. Teenagers are particularly prone to risky behaviors such as dangerous driving, drug use, and addiction. Research has shown that the teenage brain, especially the frontal lobe, develops slowly. This delayed development, coupled with the earlier maturation of the brain's reward centers, helps explain teenagers' impulsive behavior. The frontal lobe plays a crucial role in decision-making and risk assessment; its slower development means teenagers are biologically more sensitive to pleasure-seeking while having a limited capacity to fully understand potential risks. Studies indicate that the teenage brain is only about 80% developed, with some regions not reaching full maturity until around age 30[17]. This uneven developmental timeline contributes to teenagers' difficulties with planning and risk assessment. As parents and educators, it is important to educate teenagers about their brain development to help them understand their behavior, recognize unhealthy patterns, and be aware of potential warning signs of more serious issues.

Hemineglect Syndrome: Eat Only the Right Half Of the Plate

- A.B. suffered a stroke on November 5, 2007, and his life completely changed after that. Following the stroke, he began unintentionally neglecting objects, people, and sounds appearing on his left side. When walking, he veers to the right, brushing against walls and hedges, seemingly oblivious to potential hazards on his left. He is unable to go out alone due to this lack of awareness to the left. His visual neglect impacts even the simplest tasks: he shaves only the right side of his face, leaves food untouched on the left side of his plate, and paints only half of objects in his artwork, such as a pig missing half its head or a robin with details only on the right. In one test, he marked the center of a 25 cm line just 2 cm from the right edge, as if unaware of the line's left half. Even with practice, his ability to correct this has only improved slightly[18].

- Doctors found that A.B.'s stroke had damaged his right parietal lobe, a brain region at the top and back of the head responsible for controlling attention. Interestingly, A.B.'s vision remained intact; his left-side vision functioned perfectly, but his brain simply didn't "attend" to it. This condition is now known as hemineglect (also referred to as unilateral or spatial neglect), and typically arises from damage to the parietal lobe, often due to a stroke. A key characteristic of hemispatial neglect, which A.B. shows, is a lack of awareness of the left-side deficit. When asked to focus on his neglected side, he fails to understand what's missing, as his brain doesn't register the absence. This lack of awareness makes the condition particularly complex and challenging.

- Damage to the right parietal lobe is most common and results in neglect of the left side of space. This is because the right hemisphere of the brain, particularly the right parietal lobe, is responsible for attending to both sides of space. In contrast, the left parietal lobe primarily focuses on the right side of space, possibly due to its heavy involvement in language comprehension. When the right parietal lobe is damaged, the brain loses its ability to attend to the left side, while the left parietal lobe continues to attend to the right. This asymmetry explains why left-side neglect (resulting from right parietal damage) is more common and severe than right-side neglect (resulting from left parietal damage), as the intact right hemisphere can compensate by attending to both sides when the left parietal lobe is damaged.

- Patients with hemineglect can exhibit surprising and unexpected behaviors in extreme situations. For example, although they typically ignore objects or people on their left side, they may react when something significant occurs in that space. One man with hemineglect, who normally ignored his left side, consistently avoided bumping into a kitten playing near the left wheel of his wheelchair, despite never verbally acknowledging the kitten's presence[19]. Furthermore, although these patients often experience limb weakness on their left side, they can sometimes use their left limbs spontaneously. For instance, even if they struggle to lift their left arm on command, they might unconsciously use it to shoo away a fly on their face!

These actions demonstrate that while hemineglect affects conscious awareness, patients can still react to stimuli on their neglected side under certain conditions, similar to those with blindsight.

- Scientists and doctors have been exploring ways to reduce the right-side bias in hemineglect patients. One study found that motivation can improve these patients' performance. In an experiment, patients were asked to circle either coins or buttons on a cluttered page. When offered a monetary reward for circling coins during a second test, patients showed significant improvement on the coin task but not on the button task, which had no associated reward[18]. This suggests that incentives may be a useful tool in stroke rehabilitation.

Living Without A Brain

- In a remarkable 2007 Lancet study, doctors reported the unusual case of a 44-year-old French civil servant who lived a normal life despite having an extremely small amount of brain tissue. After experiencing weakness in his left leg for two weeks, the patient went to the hospital. Surprisingly, brain scans revealed a massive fluid-filled cavity occupying most of his skull, with significant enlargement of the lateral, third, and fourth ventricles. These chambers in the brain normally circulate cerebrospinal fluid, which is then drained. However, in this man's case, the fluid accumulated, causing a condition known as hydrocephalus. Despite the absence of most of his brain tissue, the man led a typical life—he was married, had two children, and had an IQ of 75, which, although below average, did not classify as a mental disability[20]. The case surprised doctors, illustrating the brain's remarkable ability to adapt and support essential life functions.

- This was by no means the first case of a patient surviving without a normal brain structure. In 1980, an article in Science titled "Is Your Brain Really Necessary?" described a remarkable case studied by neurology professor Dr. John Lorber at the University of Sheffield. A bright mathematics student, with an IQ of 126, was referred to Lorber after his campus doctor noticed his head was slightly larger than usual. A CAT scan revealed that the student had almost no brain tissue—just less than 0.04 inches (1 millimeter) of cerebral matter remained instead of the 1.77 inches (45 millimeters) typical of a normal adult[21]. Similar to the previous case, the student had hydrocephalus, a condition where cerebrospinal fluid accumulates inside the skull instead of circulating and draining properly. However, despite this condition, which is usually fatal in early childhood, the student lived a normal life and even graduated with honors in mathematics. This case challenged conventional understanding of brain function, highlighting the brain's extraordinary ability to adapt.

- The most striking case is that of Trevor Judge Waltrip, a boy born without a brain who passed away in 2014 at the age of 12. Trevor was born on Christmas Eve 2001 with hydranencephaly, a rare condition in which the cerebral hemispheres of the brain are completely absent and replaced by sacs filled with cerebrospinal fluid. He only had a brain stem, which allowed him to breathe, maintain a heartbeat, and respond to some stimuli, although he was blind and unable to verbally communicate. Despite doctors' predictions that he wouldn't survive beyond 12 weeks, Trevor defied the odds and lived for 12 years[22]. His mother noted that he was alert, disliked being alone, and was able to sense his surroundings. Trevor's condition had no known cure, and he required a feeding tube and physical therapy. The case of Trevor prompted scientists to reconsider the role of the brain in our lives, raising profound questions about the essential components of human consciousness.

Consciously Vegetative: I Know What Is Going On

- In 2024, a surprising study published in the New England Journal of Medicine revealed that at least 25% of people with severe brain injuries who appear physically unresponsive are actually conscious and aware of their surroundings. This breakthrough challenges previous assumptions about consciousness in vegetative patients and highlights the phenomenon known as cognitive motor dissociation—where individuals can respond mentally but not physically. This multicenter study involved six medical facilities in four countries: Belgium, France, the United Kingdom, and the United States. Using advanced brain imaging techniques, including functional magnetic resonance imaging (fMRI) and electroencephalography (EEG), researchers measured brain activity in response to mental tasks, such as imagining playing tennis. Despite being unable to physically respond, about a quarter of the 241 vegetative patients showed consistent brain activity during these exercises, indicating they could hear and comprehend commands[23]. The study highlights the importance of reassessing treatment strategies and ethical decisions regarding life support for patients with brain injuries. Scientists suggest that brain-computer interfaces (BCIs) could enable these patients to communicate better with their surroundings.

- This wasn't the first time scientists discovered that patients presumed to be unconscious could respond to commands and stimuli. In 2010, a study led by Dr. Adrian Owen from Cambridge University and the University of Liège in Belgium focused on a 29-year-old man who had been in a vegetative state for seven years following a traumatic brain injury from a car accident. Although he showed no outward signs of consciousness, doctors were surprised to find that he could communicate through brain activity monitored by an fMRI (functional magnetic resonance imaging) scanner. Using an innovative technique, doctors instructed the patient to think about playing tennis to indicate "yes" and to imagine walking around his home to indicate "no." The patient was able to accurately answer a series of questions about his family by generating corresponding brain activity[24]. This groundbreaking discovery revealed that the patient was not in a vegetative state but instead had a fully functioning mind trapped in an unresponsive body.

- In 2006, another study by Dr. Owen's team involved a 23-year-old woman who had been in a vegetative state for five months following a car accident. Using functional magnetic resonance imaging (fMRI), the researchers found that the woman could respond to commands by activating specific brain regions when asked to imagine playing tennis or walking through her house. Her brain activity was similar to that of healthy volunteers performing the same tasks, indicating that she could understand and follow spoken instructions despite being unable to communicate through speech or movement[25]. The researchers concluded that the woman was consciously aware of herself and her surroundings, even though she appeared to be in a vegetative state.

- These studies prompted scientists to re-examine the definition of consciousness and how it is evaluated, particularly in patients with severe brain injuries. They challenged the traditional understanding of vegetative states, suggesting that a significant number of patients might possess awareness but lack the ability to express it. Notably, Dr. Owen's team discovered that four out of 23 tested patients exhibited detectable brain activity, implying that as many as one in five patients diagnosed as vegetative could potentially be conscious. These findings have far-reaching implications for the medical field, highlighting the need for improved diagnostic tools and offering hope that these patients may one day be able to communicate their thoughts and feelings more directly.

Alice in Wonderland Syndrome: My Head Is So Big

- Alice in Wonderland Syndrome (AIWS) is a neurological condition characterized by severe distortions in a patient's perception of body size, shape, and space. These distortions are often associated with migraines, epilepsy, or other brain disorders. The syndrome, named by psychiatrist John Todd in 1955, derives its name from Lewis Carroll's *Alice's Adventures in Wonderland* due to the similarities between the condition's symptoms and the surreal size distortions described in the book. Early medical reports of similar hallucinations date back to 1904, with patients perceiving objects or body parts as significantly larger or smaller than they actually were. These hallucinations were often brief and harmless but had the potential to persist and cause disorientation.

- In one instance, a 23-year-old secretary experienced a disturbing sensation during severe migraines where her head seemed to grow to an enormous size and float upward toward the ceiling, though she remained certain it was still attached to her neck. Following these episodes, she felt unusually tall and believed she was towering over people when walking down the street, only to be shocked by her actual reflection, which showed her normal height of under five feet. This case highlights the striking disconnect between her perceived body image and reality, causing intense feelings of disorientation and fear.

- In another case, a 38-year-old housewife reported that during her migraines, parts of her body appeared to grow or distort—for example, her neck would lengthen, or her hip or flank would seem to expand. Occasionally, she felt as if she had shrunk to about a foot tall. Although she could rationalize that her perceptions were false by looking in a mirror, the vividness of the distortions was unsettling and underscored the profound disconnect between visual perception and bodily reality in AIWS.

- AIWS is a rare condition, with fewer than 200 cases reported in the medical literature since its naming in 1955. Although more common in children, with an average age of onset around nine years old, and typically associated with migraines, AIWS can also affect adults. Approximately 15% of adult migraine sufferers experience AIWS symptoms. Other reported triggers include infections like Epstein-Barr virus (which can lead to encephalitis), brain tumors, brain hemorrhages, strokes, scarlet fever, schizophrenia, and sensory deprivation. In rare cases, AIWS has been linked to hypnotherapy and altered states of consciousness, such as hypnagogic or hypnopompic states.

- AIWS remained largely obscure until recent advances in neuroimaging helped scientists investigate the relationship between brain activity and perception. Research has revealed that AIWS is a perceptual disorder linked to abnormal brain activity, particularly in regions responsible for processing sensory information, such as the parietal, occipital, and temporal lobes. Disruptions in these areas can lead to distorted visual and body perception, affecting how patients perceive their own bodies and the world around them. For example, migraines and epilepsy can alter blood flow or electrical activity in the brain, causing the misinterpretation of sensory inputs, which results in the bizarre size and spatial distortions characteristic of AIWS[26].

Anosognosia: What Do You Mean? I'm Doing Great!

- Mrs. Dodds, a stroke patient, experienced paralysis on the left side of her body due to damage in the right hemisphere of her brain. Despite this, she displayed anosognosia, a condition characterized by a lack of awareness or denial of her paralysis. Though mentally lucid and aware of her stroke, Mrs. Dodds insisted she could walk and use both hands, even though her left side remained immobile. When asked to perform tasks like touching the doctor's nose or clapping, she claimed to have completed them, though her left arm remained still. In extreme cases, patients like Mrs. Dodds might confabulate—imagining they are moving a paralyzed limb and genuinely believing they are completing tasks. When the doctor offered Mrs. Dodds a small reward for doing a one-handed task (threading a lightbulb) or a larger reward for a two-handed task (tying shoelaces), Mrs. Dodds consistently chose the two-handed task. She would then struggle with the shoelaces using only her right hand, oblivious to her limitations. When later asked, she confidently claimed to have tied the shoelaces successfully with "both hands"![8]

- Neurologists initially considered Mrs. Dodds' case to be associated with hemineglect, a condition where patients ignore the left side of their body and environment due to damage in the right hemisphere of the brain (as discussed previously). However, when doctors temporarily "paralyzed" the right, healthy arm of a denial patient named Betty, surprising results emerged. Betty was asked to move her right hand in a mirror box, while the mirror displayed the reflection of a hidden assistant's still hand. Despite seeing the still hand on her right side, Betty insisted that she saw a moving hand. This suggested that denial patients differ from those with hemineglect, which only affects the left but not the right side. Interestingly, patients with anosognosia may selectively deny paralysis but acknowledge other medical conditions, such as diabetes, indicating that anosognosia is not a global denial of all conditions but rather specific to bodily perception.

- However, doctors can sometimes use strategies to help anosognosia patients acknowledge their paralysis. For instance, in the case of a patient named Nancy, doctors addressed her denial of paralysis through an intriguing psychological experiment. Like other individuals with anosognosia, Nancy was unaware of the paralysis in her left arm. To investigate whether her denial could be overcome by modifying her belief system, the doctor conducted an informal neurological examination and pretended to inject her left arm with a "paralyzing" anesthetic (which was actually just a saline solution). After the injection, Nancy was asked if she could move her left arm, and to the doctor's surprise, she admitted that it wasn't moving, acknowledging her paralysis for the first time. To rule out the possibility that this was simply due to the doctor's persuasive influence or hypnotic suggestion, the procedure was repeated on her right (healthy) arm. When asked if she could move her right arm after the "injection," Nancy confirmed that it was functioning normally, showing no effect from the placebo injection! Interestingly, when patients like Nancy are later asked if they ever denied their paralysis, they insist that they never did, as if they have repressed their previous denial memory. This suggests that the denial associated with anosognosia may be deeply ingrained in the patient's belief system but can be temporarily influenced by external factors, such as this mock injection.

- Thanks to the advancements in neuroimaging technology, scientists and doctors now understand the brain regions affected in patients with anosognosia. Anosognosia for hemiparesis, or unawareness of paralysis, typically occurs after large lesions in the right hemisphere of the brain but can also result from smaller injuries in the right parietal area. Although it often coexists with other neglect syndromes, it is a distinct condition. Anosognosia arises from damage to the brain's ability to reconcile discrepancies in sensory input related to body image, regardless of whether the affected body part is on the left or right side. Research suggests that the left hemisphere tends to overlook these discrepancies, while the right hemisphere is more attuned to inconsistencies and disruptions. Therefore, when the right parietal area is damaged, the left hemisphere takes over and leads the patient to believe that their body image remains unchanged and healthy.

Capgras' Delusion: My Parents Are Imposters

- Arthur was attending school in Santa Barbara when he had a life-changing accident. His car collided with such force that his head struck the windshield, leaving him in a coma for three weeks. Once he regained consciousness, he underwent extensive rehabilitation, eventually recovering his speech, mobility, and memory of past events. Although he seemed mostly back to normal, he exhibited a specific delusion: he believed his parents were impostors, and nothing could convince him otherwise. He readily accepted his parents over the phone (by voice), but whenever he saw them in person, he insisted they were impostors pretending to be his parents. However, Arthur had no problem recognizing the faces of his parents or anyone else in his life, including friends, former girlfriends, and college roommates[8]. This suggested that Arthur wasn't "crazy" and that his delusion was specific rather than a sign of global mental instability.

- Arthur was diagnosed with Capgras' delusion, a rare neurological syndrome in which patients believe close relatives such as parents, spouses, children, or siblings are impostors, despite being otherwise mentally lucid. This unusual delusion can manifest in psychotic conditions such as schizophrenia but also occurs in over a third of cases involving traumatic brain injuries, like Arthur's. The syndrome was named after French psychiatrist Joseph Capgras, who, along with Jean Reboul-Lachaux, first documented it in 1923. Their study described a woman known as "Madame Macabre," who was convinced that her husband and other familiar people had been replaced by doubles.

- Examinations of Capgras patients have revealed that dysfunction in the temporal lobe, the brain region responsible for recognizing faces and objects, may contribute to the development of the delusion. When specific areas within the "what pathway" of the temporal lobe are damaged, individuals may lose the ability to recognize familiar faces, a phenomenon illustrated in Oliver Sacks' book *The Man Who Mistook His Wife for a Hat*. Typically, these recognition areas communicate with the limbic system, which generates emotional responses to familiar faces. In Arthur's case, his face recognition pathway remained intact, allowing him to identify his parents visually. However, the critical connection between this recognition area and his amygdala, the brain structure responsible for emotional responses, was damaged. As a result, Arthur could recognize his parents but experienced no emotional connection to their faces, leading him to the distressing conclusion that they were impostors.

- In 2010, scientists proposed a revised theory on Capgras syndrome to suggest a more targeted explanation. This theory posits that our brains create "hybrid representations" of people we know well, including two distinct components: one for external traits (appearance, voice) and another for internal traits (personality, emotions, preferences). According to this theory, Capgras syndrome arises when the internal component is damaged or inaccessible, causing the person to appear physically correct but feel emotionally "off," like an impostor. This explanation better aligns with patients' actual experiences and addresses drawbacks of the earlier theory, as it explains why a lack of emotional recognition alone, as observed in patients with amygdala damage (see Fearless Warrior), doesn't always lead to Capgras delusions.

The Painless Curse (A Peripheral Nervous System Anomaly But Worth Mentioning!)

- Although we typically hate pain and try our best to avoid it, some individuals never experience it at all. Edward Gibson, a vaudeville performer in the 1930s, was one such person. Despite enduring severe injuries like a blow to the face with a pickaxe, a bullet wound through his finger, a broken nose, deep cuts to his knee, and burns, he never felt pain. He could even withstand having 50-60 pins pushed into his body and once had a nail hammered through his hand without discomfort. Stefan Betz, a university student in Germany, has a similar condition that prevents him from feeling pain. He can immerse his hands in boiling water or undergo surgery without anesthesia. At the age of five, he bit off the tip of his tongue and later fractured his foot, but he experienced no discomfort. Betz couldn't remember how many times he visited the hospital. He had an infection in his left leg following a tibial bone fracture sustained while skateboarding. He has since learned to pretend to have pain to deter himself from engaging in dangerous activities and causing further harm[27].

- Edward Gibson and Stefan Betz are among the few hundred people in the world who suffer from a condition known as congenital insensitivity to pain, or CIP. It is a rare genetic disorder and was first documented in 1932. People with CIP retain normal sensory responses, such as temperature perception, but often engage in self-destructive behaviors that can lead to premature death. For example, a Pakistani boy became famous for walking on hot coals and stabbing himself without pain but tragically died after jumping off a roof. Many CIP patients face significant risks, including joint damage, infections, and accidental death.

- Scientists have discovered that CIP patients carry mutations in the SCN9A gene. These mutations disrupt the Nav1.7 sodium channel, which is responsible for pain perception. This discovery has sparked significant interest in developing drugs that target this sodium channel, offering potential relief for patients with conditions like neuropathic pain and osteoarthritis. In addition, researchers have identified the PRDM12 gene, which regulates pain neuron activity, as another promising target for pain therapies. Although CIP patients may not directly benefit from these advancements, their participation in research provides crucial insights into pain mechanisms and the development of pain management medications.

6.5 Brain Enhancement with Artificial Intelligence (AI): Friend or Foe?

Become a Superhuman: Empowered by AI

- Can we achieve speechless communication? AI is making it possible. In 2018, MIT researchers developed a computer interface called the AlterEgo project, which can transcribe words that users "speak silently" within their minds. The system utilizes a wearable device with electrodes placed on the face and jaw to detect subtle neuromuscular signals generated by internal verbalizations. These signals are then processed by a machine-learning algorithm that matches them to specific words[28]. This technology allows users to interact with computational tasks without speaking or using a visible interface. Additionally, the device incorporates bone-conduction headphones, which transmit sound through facial bones, allowing users to receive information discreetly without obstructing their hearing. A key application demonstrated with the AlterEgo project was playing a silent chess game, where users could communicate their moves and receive computer recommendations without vocalizing them. The ultimate goal is to create a seamless intelligence-augmentation platform that integrates human cognition with computational power. Current prototypes require only four electrodes, and studies have shown a 92% transcription accuracy rate. This technology holds immense potential for applications in noisy environments, scenarios requiring silent communication (e.g., special operations), and assisting individuals with speech disabilities, such as those who have lost their vocal capabilities.

- The power of AI doesn't stop there. In 2024, MIT's Fluid Interfaces research group, led by Flemish scientist Pattie Maes, is developing a brain-computer interface (BCI) that allows users to control objects using their minds. The device, resembling a pair of glasses, uses sensors to read brainwave activity through electroencephalogram (EEG) and track eye movements. This technology can determine a user's focus and attention, potentially aiding in the treatment of conditions like attention-deficit/hyperactivity disorder (ADHD) by providing feedback on attentiveness. More strikingly, it holds promise for individuals with severe disabilities, such as advanced amyotrophic lateral sclerosis (ALS), by enabling them to communicate through their brain signals. For instance, patients could look at letters to spell words, which would then be vocalized by an app[29]. Maes, a pioneer in Human Augmentation, prioritizes enhancing human intelligence over developing machines that surpass it. She advocates for using technology to augment human abilities, empowering people to achieve what is currently beyond their reach. This vision contrasts with the approach of companies like OpenAI and Google, which focus on creating intelligent machines. Maes believes that human augmentation is a more compelling path, aiming to transform individuals into "superhumans" rather than having AI replace human roles.

- Meanwhile, AI also has tremendous potential in advancing our healthcare. Neuralink, a leading developer of brain-computer interface (BCI) technology, recently showcased its progress with a video featuring the first human recipient of its BCI device. The recipient, a 29-year-old paralyzed from the shoulders down, demonstrated the ability to play computer chess by controlling a cursor with his mind—a feat similar to experiments from 2006 but achieved with significantly more sophisticated technology. The implant is wireless and unobtrusive, consisting of a coin-sized hub in the skull and 64 flexible threads inserted into the motor cortex to record neural activity. Although initial challenges with electrode stability required adjustments, the updated design reflects progress toward seamless brain-computer integration[30]. Neuralink, driven by founder Elon Musk's ambition to both enhance and restore brain functions, has outlined goals for future BCIs to address conditions like psychosis, seizures, and memory loss, with potential applications involving AI integration. Despite these advancements, experts urge caution, emphasizing the need to address ethical considerations before broader human application.

- While Neuralink may dominate headlines in the brain-computer interface field, several other companies are making significant strides with diverse approaches.

New York-based Synchron, for instance, has already implanted its device in ten individuals and launched a patient registry for a larger clinical trial. The company is backed by high-profile investors like Bill Gates and Jeff Bezos. Although most companies in this field aim to assist those unable to move or speak by capturing brain activity to interpret user intentions, their methods vary in invasiveness. Neuralink and Paradromics, for example, opt for deep electrode insertion into the cortex. This allows for high-resolution data capture, enabling tasks like speech decoding. In contrast, Synchron's "Stentrode" bypasses open brain surgery by placing electrodes in blood vessels above the motor cortex. Meanwhile, Blackrock Neurotech's Utah array, a staple in academic research since 2004, has received an FDA Breakthrough Designation. However, it offers lower data bandwidth than Neuralink's device. Another emerging player, Precision Neuroscience, employs a minimally invasive method with a thin electrode array that slides over the cortex. Motif Neurotech, on the other hand, is pursuing an innovative non-invasive device, focusing on treating mood disorders rather than decoding complex data. With such rapid advancements, controlling technology via thought is no longer science fiction, though the broader transformative goals set by companies like Neuralink remain distant.

AI Concerns: The Risk of Domination and Displacement

- In March 2016, the AI algorithm AlphaGo, developed by Google's DeepMind, achieved a groundbreaking victory by defeating the world's best Go player Lee Sedol 4-1. Mastering the ancient game of Go has long been considered one of the toughest challenges for artificial intelligence. Unlike previous AI successes in games like chess, which relied on preprogrammed strategies, AlphaGo learned to play Go using a general-purpose deep-learning algorithm. This approach allowed AlphaGo to interpret the complex patterns and strategies in Go, which involves more possible board configurations than there are atoms in the universe. AlphaGo's remarkable performance is highlighted by its ability to defeat European Go champion Fan Hui five times in a row and its dominance over other computer programs, winning 99.8% of games. What makes this even more extraordinary is that AlphaGo plays in a way that resembles human intuition and strategy, rather than relying purely on brute computational force[31]. Such human-like behavior highlights the power of deep learning and its potential applications in complex problem-solving beyond gaming, such as medical diagnoses and climate modeling.

- Despite its convenience, artificial intelligence poses growing issues. One such issue is deepfake AI scams targeting reputable people, such as scientists, with significant reputational and professional risks. For example, South African urologist Kgomotso Mathabe discovered a deepfake video of herself promoting a fake drug, leading to financial scams and reputational damage. Similarly, Indian diabetes specialist Viswanathan Mohan faced deepfake attacks, with videos falsely depicting him endorsing herbal products[32]. These AI-generated videos caused a lot of social and ethical problems.

- The rapid rise of AI is also reshaping the job market, creating both opportunities and disruptions. A 2023 ResumeBuilder survey indicated that 37% of companies utilizing AI have already replaced workers, and 44% anticipate AI-driven layoffs in 2024. This trend is impacting a wide range of industries, including healthcare, agriculture, and manufacturing. Many experts agree that AI will fully automate certain jobs within the next five to ten years. Roles particularly at risk include customer service representatives, drivers, computer programmers, research analysts, paralegals, factory workers, financial traders, travel advisors, content writers (in some cases), and graphic designers. However, certain jobs are less likely to be replaced by AI due to their inherently creative, empathetic, or complex nature. These include roles such as teachers, nurses, social workers, therapists, handypersons, lawyers, HR specialists, writers, and artists[33].

- Despite concerns about automation, AI offers significant advantages in the workplace. These include a focus on more engaging tasks, improved efficiency and potentially shorter workdays, more informed decision-making, enhanced innovation and problem-solving capabilities, and the ability to provide more personalized customer interactions. Additionally, AI is creating new jobs, particularly in machine learning and AI ethics, as human oversight is essential for training, monitoring, and refining AI systems. The impact of AI can be likened to the Industrial Revolution, where job roles shifted and evolved rather than simply disappearing, leading to new types of work opportunities. The MIT Shaping the Future of Work Initiative, launched on January 22, 2024, and led by professors Daron Acemoglu, David Autor, and Simon Johnson, proposes developing AI tools that augment human workers rather than replace them. This approach aims to empower less-skilled workers to take on more specialized tasks and create a wider range of job opportunities for individuals with diverse skill sets.

Notes

1. Walløe, S., Pakkenberg, B., & Fabricius, K. (2014). Stereological estimation of total cell numbers in the human cerebral and cerebellar cortex. *Frontiers in Human Neuroscience, 8*. https://doi.org/10.3389/fnhum.2014.00508

2. Solan, M. (2021, August 1). *The book of neurogenesis*. Harvard Health. https://www.health.harvard.edu/mind-and-mood/the-book-of-neurogenesis

3. Spector, R. (2022, September 21). *How men's and women's brains are different | Stanford Medicine*. Stanford Medicine Magazine. https://stanmed.stanford.edu/how-mens-and-womens-brains-are-different/

4. Banissy, M. J., & Ward, J. (2007). Mirror-touch synesthesia is linked with empathy. *Nature Neuroscience, 10*(7), 815–816. https://doi.org/10.1038/nn1926

5. Bates, B. C. (2018, April 3). *The doctor who really feels his patients' pain*. https://www.bbc.com/news/stories-43572639

6. Feinstein, J. S., Adolphs, R., Damasio, A., & Tranel, D. (2010). The human amygdala and the induction and experience of fear. *Current Biology, 21*(1), 34–38. https://doi.org/10.1016/j.cub.2010.11.042

7. Feinstein, J. S., Buzza, C., Hurlemann, R., Follmer, R. L., Dahdaleh, N. S., Coryell, W. H., Welsh, M. J., Tranel, D., & Wemmie, J. A. (2013). Fear and panic in humans with bilateral amygdala damage. *Nature Neuroscience, 16*(3), 270–272. https://doi.org/10.1038/nn.3323

8. Ramachandran, V. S., & Blakeslee, S. (1999). *Phantoms in the brain*. Fourth Estate.

9. De Haan, E. H. F., Corballis, P. M., Hillyard, S. A., Marzi, C. A., Seth, A., Lamme, V. a. F., Volz, L., Fabri, M., Schechter, E., Bayne, T., Corballis, M., & Pinto, Y. (2020). Split-Brain: What We Know Now and Why This is Important for Understanding Consciousness. *Neuropsychology Review, 30*(2), 224–233. https://doi.org/10.1007/s11065-020-09439-3

10. Robson, D. (2022, February 24). *Blindsight: the strangest form of consciousness*. https://www.bbc.com/future/article/20150925-blindsight-the-strangest-form-of-consciousness

11. De Gelder, B., Tamietto, M., Van Boxtel, G., Goebel, R., Sahraie, A., Van Den Stock, J., Stienen, B. M., Weiskrantz, L., & Pegna, A. (2008). Intact navigation

skills after bilateral loss of striate cortex. *Current Biology, 18*(24), R1128–R1129. https://doi.org/10.1016/j.cub.2008.11.002

12. Kim, H., Han, J., Park, Y. H., Kim, B. J., Yang, W., & Kim, S. (2017). Supernumerary phantom limb in a patient with basal ganglia hemorrhage - a case report and review of the literature. *BMC Neurology, 17*(1). https://doi.org/10.1186/s12883-017-0962-7

13. Zernitz, M., Rizzello, C., Rigoni, M., & Van De Winckel, A. (2024). Case Report: Phantom limb pain relief after cognitive multisensory rehabilitation. *Frontiers in Pain Research, 5*. https://doi.org/10.3389/fpain.2024.1374141

14. ABC News. (2011, October 7). *Ex-Football player's life is deleted by retrograde amnesia*. https://abcnews.go.com/Health/football-player-scott-bolzans-life-deleted-irreversible-amnesia/story?id=14616045

15. *Researchers find "lost" memories*. (2015, May 28). MIT News | Massachusetts Institute of Technology. https://news.mit.edu/2015/optogenetics-find-lost-memories-0528

16. Day, E. (2017, February 22). He was bad, so they put an ice pick in his brain. . . *The Guardian*. https://www.theguardian.com/science/2008/jan/13/neuroscience.medicalscience

17. Edwards, S. (2011, December 19). Deciphering the teenage brain. *Harvard Medical School*. https://hms.harvard.edu/news/deciphering-teenage-brain

18. Hoffman, T. (2012, November 23). The man whose brain ignores one half of his world. *The Guardian*. https://www.theguardian.com/science/blog/2012/nov/23/man-brain-ignores-half-world

19. Ogden, J. (2012, September 11). The bizarre disorder of hemineglect | BrainLine. *BrainLine*. https://www.brainline.org/article/bizarre-disorder-hemineglect

20. Andrew, E. (2021, September 24). *Meet the healthy, functioning man who survived with almost no brain*. IFLScience. https://www.iflscience.com/man-tiny-brain-lived-normal-life-31083

21. Lewin, R. (1980). Is your brain really necessary? *Science, 210*(4475), 1232–1234. https://doi.org/10.1126/science.7434023

22. Ksla. (2014, September 12). Keithville boy born without brain dies at 12. *https://www.ksla.com*. https://www.ksla.com/story/26405843/keithville-boy-born-without-brain-dies-at-12/?autostart=true

23. Bodien, Y. G., Allanson, J., Cardone, P., Bonhomme, A., Carmona, J., Chatelle, C., Chennu, S., Conte, M., Dehaene, S., Finoia, P., Heinonen, G., Hersh, J. E., Kamau, E., Lawrence, P. K., Lupson, V. C., Meydan, A., Rohaut, B., Sanders, W. R., Sitt, J. D., . . . Schiff, N. D. (2024). Cognitive motor dissociation in disorders of consciousness. *New England Journal of Medicine*, *391*(7), 598–608. https://doi.org/10.1056/nejmoa2400645

24. Boseley, S. (2017, November 26). Think tennis for yes, home for no: how doctors helped man in vegetative state. *The Guardian*. https://www.theguardian.com/science/2010/feb/03/vegetative-state-patient-communication

25. Owen, A. M., Coleman, M. R., Boly, M., Davis, M. H., Laureys, S., & Pickard, J. D. (2006). Detecting awareness in the vegetative state. *Science*, *313*(5792), 1402. https://doi.org/10.1126/science.1130197

26. The MIT Press Reader. (2022, October 3). *The curious case of Alice in Wonderland syndrome*. https://thereader.mitpress.mit.edu/the-curious-case-of-alice-in-wonderland-syndrome/

27. Cox, D. (2022, February 24). *The curse of the people who never feel pain*. https://www.bbc.com/future/article/20170426-the-people-who-never-feel-any-pain

28. *Computer system transcribes words users "speak silently."* (2018, April 4). MIT News | Massachusetts Institute of Technology. https://news.mit.edu/2018/computer-system-transcribes-words-users-speak-silently-0404)

29. Flanders News. (2024, April 5). Startling results: 'Brain-computer interface' allows user to move objects with mind (VIDEO). *vrtnws.be*. https://www.vrt.be/vrtnws/en/2024/04/05/augmented-reality-ai-mit/)

30. Naddaf, M., & Drew, L. (2024). Second brain implant by Elon Musk's Neuralink: will it fare better than the first? *Nature*, *632*(8025), 481–482. https://doi.org/10.1038/d41586-024-02368-8

31. Gibney, E. Google AI algorithm masters ancient game of Go. *Nature* 529, 445–446 (2016). https://doi.org/10.1038/529445a

32. Nordling, L. (2024). Scientists are falling victim to deepfake AI video scams — here's how to fight back. *Nature*. https://doi.org/10.1038/d41586-024-02521-3

33. Urwin, M. (2024, July 19). AI taking over jobs: What to know about the future of jobs. *Built In*. https://builtin.com/artificial-intelligence/ai-replacing-jobs-creating-jobs

Chapter 7

The Elusive Yet Powerful Mind

7.1 The Unbelievable Mind-Body Connection

The Placebo Effect: Feel It When You Believe It

- Our brain can be tricked into believing a fake treatment is real, triggering a healing response known as the placebo effect. Scientific research has demonstrated that placebos can be as effective as conventional treatments in relieving pain for migraine and osteoarthritis. Placebos likely work by triggering the release of neurotransmitters such as endorphins and dopamine, which are associated with pleasure and well-being.

- Placebos often work because people don't know they're taking one. But what if you do know? A study published in *Science Translational Medicine* tested three different treatments in patients suffering from migraine attacks: one group received a migraine drug labeled with the drug's name, another received a placebo labeled "placebo," and a third group received no drugs or placebos. Researchers found that the placebo was 50% as effective as the real drug in reducing migraine pain, even when patients knew they were taking a placebo[1].

- Contrary to the placebo effect is the nocebo effect. The nocebo effect describes negative outcomes that can occur if we believe something will harm our body. Research shows that if we believe a treatment will be painful or have side effects, we are more likely to experience pain or side effects (such as muscle spasms and chest pain). Common nocebo effects include pain, muscle aches, sexual side effects, itching, and memory problems.

- The nocebo effect seems to be more common in people who experience anxiety or depression, with certain personality traits (such as being pessimistic, neurotic, or highly competitive), or who are told that the treatment will cause pain or unwanted side effects. In a 2017 study, scientists found that the price of a medication can modulate pain processing. Patients reported stronger pain when a fake treatment was labeled as an expensive medication compared to when it was labeled as a cheap medication, even though they received no active treatment[2].

Counterclockwise: Aging Is All in Your Mind

- Probably the most striking experiment about the mind-body connection was the "Counterclockwise" experiment done by Harvard social psychologist Ellen Langer. In 1979, Langer invited eight elderly men in their late 70s and early 80s to a retreat meticulously designed to resemble 1959—transporting them 20 years into the past. The environment was carefully recreated with period-specific furniture, newspapers, books, and television shows. The men were instructed to fully immerse themselves in the experience as if they were actually back in 1959. This involved discussing current events from that era, speaking about themselves in the present tense as if they were younger, and avoiding any references to their actual age. The results were remarkable: after just one week, the participants showed measurable improvements in several health indicators, including vision, hearing, memory, dexterity, and strength. Their cognitive performance also improved. Moreover, when photos taken before and after the retreat were shown to independent observers, the men appeared noticeably younger in the "after" images[3].

- This mind-body connection isn't limited to the elderly. In 2016, Langer's research team discovered that our minds can influence blood sugar levels simply by believing how much time has passed. In this study, researchers randomly assigned 47 participants with type 2 diabetes to one of three time perception conditions: Fast, Normal, and Slow. Participants were then instructed to play simple video games for 90 minutes. Those in the Normal condition saw a clock displaying the actual elapsed time, while those in the Slow and Fast conditions saw clocks that indicated 45 and 180 minutes had passed, respectively. Surprisingly, at the end of the 90-minute session, participants' blood glucose levels reflected their perceived time (45, 90, or 180 minutes) rather than the actual elapsed time[4]. This suggests that a simple mental trick can lead the body to adjust its blood glucose levels.

- In a similar 2023 study, Langer's team inflicted minor wounds on participants and used the same three time perception conditions, with the actual elapsed time being 28 minutes for all groups. The wounds healed significantly faster in the Fast Time condition compared to the Normal and Slow Time conditions, further supporting the concept of mind-body unity[5].

Mental Exercises: Get Stronger & Slimmer... Using Your Mind?

- If you still believe that weight loss is only achieved by burning a specific number of calories, think again. A fascinating study investigated how beliefs about physical activity can influence health, even without actual behavioral changes. In this study, researchers recruited 84 hotel housekeepers, informing half that their daily work (cleaning rooms) met federal exercise guidelines, while the other half remained uninformed. Surprisingly, after just four weeks, the informed group exhibited significant physical improvements: weight loss, lower blood pressure, improved waist-to-hip ratios, and reduced body mass indexes—all without changes in activity, workload, or diet[6]. These findings support the concept of "Activity Adequacy Mindsets" (AAMs), suggesting that believing one's activity level is healthy can positively impact physiological measures like heart rate and blood pressure[7].

- In another experiment, scientists found that mental training can induce strength gains without performing physical exercises. In their experiment, 30 young, healthy volunteers were divided into four groups to train their little finger abductor and elbow flexor muscles, which are frequently used in daily activities. One group of 8 people performed "mental contractions" of little finger abduction, one group of 8 performed mental contractions of elbow flexion, one group of 6 underwent physical muscle exercises for little finger abduction and elbow flexion, and the last group of 8 did not perform physical or mental training but just participated in all experiments as a control group. After 12 weeks of training 15 minutes per day, 5 days per week, the physical training group increased their finger abduction strength by 53%. Interestingly, the mental training groups had 35% and 13.5% increases in their finger abduction and elbow flexion strength, respectively, while the control group did not show a significant strength gain in their muscles. Scientists confirmed the muscle strength improvement of mental training groups by measuring the electroencephalogram-derived cortical potential[8]. Maybe try it out next time to grow your muscles when you feel too tired to go to the gym!

- The above experiments are not isolated cases. Since the 1930s, scientists have been studying how mental practices can improve athletic performance. In an earlier study in 1978, 32 college students were randomly assigned to one of four groups to test the impact of mental practice on dart-throwing ability. The control group received no practice, while the other three groups engaged in different forms of practice: mental rehearsal only, mental rehearsal with simulated motor movements, and direct physical practice. Participants in the mental rehearsal group were instructed to close their eyes and vividly imagine the process of throwing darts, focusing on sensory details like the feeling of the darts in their hands, the sight of the target, and the sound of the darts hitting the board. The mental/motor group combined mental rehearsal with physical motions, mimicking the act of throwing without an actual dart. The direct practice group engaged in traditional physical practice, throwing darts at the target. After six days, all practice groups showed significant improvement compared to the control group. Remarkably, the mental practice group improved almost as much as the direct physical practice group[9], highlighting again the power of the mind!

- The dart-throwing ability is only one of the many motor skills tested. In fact, scientists have extensively studied the effects of mental practice on various motor skills, including basketball free throws, badminton, speed skating, tap dancing, wand juggling, softball pitching, golf chipping, and soccer dribbling. Through meta-analysis (a statistical technique that combines the results of multiple studies), researchers have concluded that mental practice significantly improves motor skill learning and retention[10,11]. Due to its positive impact, mental practice is now utilized in various fields. Athletes use it to enhance performance in sports like basketball, golf, and dance. Musicians, such as pianists, employ mental practice to learn new pieces, while medical professionals, including surgeons, refine their procedural skills through visualization. Additionally, mental practice aids in stroke recovery by helping patients regain mobility and can reduce strength loss in immobilized limbs due to injury.

The Healing Power of Mind

- The healing journey of Tibetan lama Phakyab Rinpoche offers a remarkable example of the mind's potential to heal the body. The author of the book *Meditation Saved My Life*, Rinpoche faced severe medical conditions after enduring brutal imprisonment in China and a grueling escape. He developed gangrene in his right ankle, a life-threatening condition that occurs when tissue dies due to a lack of blood flow or a serious bacterial infection. Additionally, he suffered from shattered bones and spinal tuberculosis. Doctors in New York believed amputation was the only way to save his life. However, a letter from the Dalai Lama, posing the question, "Why do you seek healing outside of yourself?", inspired Rinpoche to explore an alternative path: meditation. With years of meditation experience and a deep understanding of Tibetan healing practices, Rinpoche left the hospital and embarked on a self-guided retreat. He dedicated fourteen hours daily to meditation and prayer, gradually defying medical expectations. His infected ankle healed, and his bones began to regenerate. Three years later, he achieved a full recovery, regaining the ability to walk normally.

- In an interview, Phakyab Rinpoche emphasized that compassion was key to his healing success. He believes that true healing begins in the mind and is available to everyone. It's not just about medical miracles; it's about cultivating compassion and inner transformation in daily life. Rinpoche approaches others with an open heart, seeing them as fellow human beings, each with the potential for kindness and growth. He emphasizes the practice of compassion and loving-kindness meditation for cultivating inner peace. According to Rinpoche, excessive self-focus can hinder peace, while compassion opens the "inner door." This quiets self-centered thoughts and fosters openness in the heart and mind. Meditation, particularly "shamatha" (calm-abiding meditation), silences mental distractions, promoting a sense of calm in both body and mind[12].

- Using advanced neuroimaging techniques, neuroscientists have revealed how meditation can reshape our brains. A Harvard research team led by Sara Lazar utilized functional magnetic resonance imaging (fMRI) scans to examine 16 participants before and after an eight-week mindfulness program. Participants engaged in mindfulness exercises for an average of 27 minutes each day. Following the eight-week program, all participants reported significant improvements on mindfulness questionnaires. Brain scans revealed increased gray matter in the hippocampus, a brain region crucial for learning and memory. In addition, participants also reported reduced stress levels, which correlated with decreased gray matter density in the amygdala, an area associated with anxiety and stress[13]. These brain changes were observed exclusively in the meditation group, suggesting that mindfulness practice can trigger structural and functional changes in our brains.

Finger Growing: A 1-Minute Experiment to Test Your Mind's Power

- After reading all this, have you wondered about your own mental power? Now, here's your chance. Your mind can change your body in seconds, and your fingers are a great target for your mind's power. To grow your fingers using your mind, start by aligning your left and right hands to compare the length of your fingers—they should be similar, though not exactly the same. Next, take your hands apart, close your eyes, and relax. Stick out the hand you want to grow and focus on it, repeating to yourself, "It is growing longer, it is growing longer." Visualize something that grows, like grass sprouting or a tree shooting up, to help reinforce the idea. After practicing this for 15-30 seconds (it can take longer to see an effect), open your eyes, realign your hands, and compare your fingers to see how much longer they've grown. I expect the result will surprise you!

7.2 The Unconscious Mind: Operating Beyond Awareness

- Sigmund Freud, often referred to as the "father of psychology," developed psychoanalysis theory in the early 1890s. This theory significantly impacted our understanding of the human mind and its functions. Freud proposed that the human mind consists of three levels of consciousness: the conscious, the preconscious (or subconscious), and the unconscious. He emphasized the substantial role of the unconscious mind in influencing behavior. The unconscious mind, according to Freud, houses repressed memories, thoughts, and urges that are not easily accessible to our conscious awareness. These three levels of consciousness are often compared to an iceberg, where the conscious mind is merely the visible tip above the water, while the unconscious mind represents the vast, hidden portion beneath the surface.

- While we use our conscious minds daily, we often don't realize that many of our behaviors are driven by the unconscious. This is vividly illustrated in cases of split-brain patients and those with blindsight (Chapter 6). For example, split-brain patients might accurately point or draw something but be unable to verbally explain their actions. Similarly, individuals with blindsight can navigate a hallway and avoid obstacles perfectly, all without conscious awareness of how they do it. In an extreme example, patients with anosognosia seemed to suppress memories into their unconscious mind and switch between two distinct identities: one believing they were perfectly healthy, the other acknowledging their health conditions. These phenomena raise a profound question: Is our free will—the conscious mind we rely on—truly free, or is it merely an illusion?

- Perhaps the most direct evidence of our subconscious/unconscious mind's role comes from brain scan studies. Using functional magnetic resonance imaging (fMRI), scientists can observe subjects' brain activity during decision-making. In one experiment, subjects watched a stream of letters on a screen and, whenever they felt the urge, chose to press either a left or right button. They were asked to remember the letter displayed at the moment of decision and later indicated their choice by selecting that letter. Researchers used fMRI to analyze brain activity to see if they could predict each subject's choice before it reached conscious awareness. They found that two brain regions—the frontopolar cortex (BA10) and the parietal cortex—showed activation that predicted the button press up to 10 seconds before the subject consciously decided[14]. This suggests that unconscious brain activity influences decisions even before they enter our conscious awareness. A subsequent study on consumer product selection further revealed that brief periods of unconscious thought (UT) within specific brain regions (namely, the right dorsolateral prefrontal cortex and left visual cortex) helped improve decision-making. This implies that our brains can process and integrate information to support smarter decisions, even without conscious thought[15].

- Scientific evidence suggests that our unconscious mind has at least two properties. First, it is very powerful and can change our bodies in ways the conscious mind cannot. While our conscious mind can temporarily influence our bodies—like lengthening our hands/fingers through visualization—these effects last only a few minutes. In contrast, the unconscious mind could produce lasting changes. For example, studies have shown that patients can eliminate warts—hard, raised bumps caused by the human papillomavirus (HPV)—simply by "thinking them away." In a study published in *The Lancet*, 21 patients with warts underwent hypnosis involving relaxation, eye fixation, and repetitive commands. During hypnosis, they were given two suggestions: that warts on one side of their body would disappear, and that they would perform a specific action upon waking. Of the 21 patients, 15 achieved a sufficient hypnotic depth, and their warts disappeared within five weeks to three months[16]. These surprising results, confirmed by multiple other studies[17,18,19], highlight a connection between depth of hypnosis, post-hypnotic suggestion, and wart disappearance.

- Second, unlike the conscious mind, our unconscious mind is constantly aware of our surroundings, even when we are unconscious. In Brian Weiss's book "Many Lives, Many Masters," a 27-year-old lab technician named Catherine sought treatment from Weiss for severe anxiety and panic attacks. Weiss used regression therapy, and under hypnosis, Catherine recalled many of her 86 previous lives. In one therapy session, Catherine was able to recall details of the doctors' conversations about her throat condition during vocal cord surgery. Although anesthetized and deeply unconscious at the time, her mind had registered the surgeons' discussion about the risk of choking due to the insertion and removal of a tube. This awareness from her unconscious mind triggered an intense fear of swallowing or choking when she woke up. During the hypnosis session, Catherine realized this connection and, as a result, overcame her anxiety and fear[20]. Our unconscious mind likely collects a vast amount of information while we are in an unconscious state (like sleep). The only obstacle is accessing this information with our conscious mind.

7.3 The Quantum Puzzle of the Mind

- While neuroscientists and psychologists still debate the reality of mind-body unity, physicists have embraced an even bolder concept: the universe itself might be a projection of our minds. In 2005, to celebrate the centennial of Einstein's "miracle year," Nature, a prestigious science journal, published "The Mental Universe," a concise essay by Richard Henry, a Professor of Physics and Astronomy at Johns Hopkins University. Henry's essay opens with a provocative statement: "The only reality is mind and observations, but observations are not of things. To see the Universe as it really is, we must abandon our tendency to conceptualize observations as things."[21] Many other physicists, including Amit Goswami, a retired professor of Theoretical Science at the University of Oregon and author of "Physics of the Soul," share similar perspectives. This view is intriguing, but what leads physicists to such a radical conclusion about the mind?
- This is likely due to a strange phenomenon in quantum mechanics called the "observer effect," where the act of observation changes reality. The classic example

is the double-slit experiment: When not observed, a single electron seems to pass through both slits simultaneously (a bizarre quantum phenomenon known as superposition), creating an interference pattern on a screen. However, if scientists try to observe which slit an electron passes through, the electron behaves differently, acting like a particle instead of a wave. This causes the interference pattern to disappear[22]. In other words, it's as if electrons know they're being watched and change their behavior! Since we can't help but observe the world as we perceive it, scientists are left wondering: Is our observable universe truly real? Does an objective reality even exist?

- Schrödinger's cat may help us better appreciate this puzzling concept. In this thought experiment, the Austrian physicist Erwin Schrödinger proposed that a hypothetical cat's fate depends on a random event with a 50% chance of occurring, such as radioactive decay. Until someone opens the box to measure (observe) the decay, it remains in a superposed state—that is, both decayed and not decayed, with each at a 50% chance. As a result, the cat is, paradoxically, both alive and dead simultaneously until the box is opened and the cat's state is measured by an outside observer. This may sound ridiculous, but in quantum theory it is completely rational. In other words, we are constantly observing the universe, and whenever we do so, we are creating our own reality with our observations and minds.

- To resolve the paradox of Schrödinger's cat, US physicist Hugh Everett III developed the many-worlds (or parallel universes) theory in the late 1950s. This theory proposes that every quantum event causes a split in reality, creating parallel universes that contain every possible outcome. This implies that Schrödinger's cat could exist in two parallel universes: alive in one and dead in another. The cat's dead and alive states are decoherent, meaning they do not interact or communicate with each other. Once a measurement (observation) is made, the observer and the cat exist in only one of these worlds, not both. So, if you constantly regret making wrong choices in your life, take solace in the thought that another version of you might be enjoying a happy ending in a parallel universe!

- By the way, it's worth noting that despite seeming strange and counterintuitive, quantum physics is widely regarded as the most successful scientific theory ever developed. It explains a vast range of natural phenomena with extraordinary accuracy, particularly at the atomic and subatomic levels. Additionally, it explains phenomena ranging from the behavior of atoms to the properties of semiconductors and the Sun. Many modern technologies, such as lasers and transistors, rely on quantum principles, and quantum physics has paved the way for groundbreaking innovations like computers and digital cameras.

7.4 Mind and the Universe: Not Separate but Interconnected

Hearts Beating Together

- As human beings, our bodies may seem separate, but they are intricately connected. In the Spanish village of San Pedro Manrique, locals participate in a fire-walking ritual every June 23, crossing hot coals in a ceremonial celebration of the summer solstice. To understand the physiological impact of such rituals, researchers conducted a study focusing on heart rate. The team monitored 12 fire-walkers, 9 of their relatives, and 17 visiting spectators. Interestingly, the researchers found that the heart rates of observers who were relatives or close friends of the fire-walking participants synchronized with those of the fire-walkers. These relatives' heart rates spiked and dropped almost identically to those of the fire-walkers before, during, and after each walk, even when they were not actively participating themselves[23]. This synchrony lasted for the entire 30-minute ritual, suggesting a deep physiological connection tied to emotional bonds rather than shared physical actions. In contrast, visiting spectators who had no personal connection to the fire-walkers did not show the same heart rate pattern, underscoring that close social connections are likely key to this synchronized response.

- Have you ever felt a deep connection with your baby while sharing a smile? Scientists have discovered the underlying mechanism behind this heartwarming experience. A 2011 study reveals that the hearts of mothers and their 3-month-old infants can synchronize down to the millisecond, simply by sharing smiles and joyful interactions. Researchers observed 40 mother-infant pairs during face-to-face interactions, using electrodes to monitor their heartbeats. They found that the mother's and child's hearts beat in near-perfect synchrony during moments of warm eye contact or cooing sounds[24]. Notably, this effect was only observed between each mother and her own child. This "cardiac coupling" appears to activate specific brain areas associated with empathy and social connection. Previous research suggests that babies who miss out on these bonding moments, such as those born prematurely or whose mothers experience postpartum depression, may exhibit reduced empathy later in life, during their teenage years.

- A similar physiological connection has been observed in couples. In 2013, a UC Davis study found that romantic partners can synchronize their heart rates and breathing patterns, supporting the notion that their hearts truly "beat as one." Researchers, led by psychology professor Emilio Ferrer, monitored 32 heterosexual couples who were seated a few feet apart quietly without speaking or touching. The study found that both heart rate and breathing intervals aligned closely within each couple. However, this synchronization did not occur when individuals were paired with non-partners[25]. The study also revealed that women tended to adjust their physiological responses more to match their partners, suggesting a possible heightened level of empathy or attunement in women.

- In the scenarios described above, the couples did not touch each other. But what if they held hands? A subsequent study by CU Boulder researcher Pavel Goldstein in 2017 explored how touch can alleviate pain in romantic couples through physiological synchronization. In a study of 22 long-term heterosexual couples, researchers measured heart and breathing rates in three scenarios: sitting together without touching, sitting together holding hands, and sitting apart while the woman experienced mild pain. Similar to the 2013 study, scientists found that couples naturally synchronize their physiological rhythms, but pain disrupts this synchronization. However, when the couples held hands, their heart and breathing rates synced again, and the woman's pain decreased[26]. The pain-relieving effect was stronger when the male partner exhibited higher levels of empathy, suggesting that touch may communicate empathy and create an analgesic effect. Researchers hypothesize that synchronization may influence the brain's anterior cingulate cortex, an area associated with both empathy and pain perception. So, the next time you're in the delivery room, remember the power of a supportive touch—hold your partner's hand!

Calming the Mind, Calming the City

- If the observable universe is a projection of our minds, then changing our mental state could not only change our bodies (through the mind-body connection) but also impact the universe, including our living environment. In fact, many research groups have studied how meditation practices impact social crime rates, and they've found an interesting connection between the two. Below are two notable studies.

- In one study, researchers used time series analysis to examine the impact of Transcendental Meditation (TM) on the crime rate in Merseyside, United Kingdom. Merseyside is a metropolitan county in North West England that includes the city of Liverpool (population 1.5 million) and has a history of high crime rates. In 1975, a Liverpool TM instructor named Patricia Stone proposed a plan to the Lord Mayor to reduce crime by 16% through Transcendental Meditation. Previous research suggested that when the number of meditation practitioners in a community reached the square root of 1% of the population, a measurable shift toward greater social order (indicated by reduced crime) could occur. In Merseyside's case, this threshold was the square root of roughly 10,000, which equals 100 practitioners. With financial support from the Mayor, Mrs. Stone aimed to teach TM to 1% of the population. In 1980, Maharishi European Sidhaland was established in nearby Skelmersdale to promote an ideal lifestyle through daily TM-Sidhi practices. By 1991, this community had grown to 365 members. In addition, a group meditation center called the Maharishi Golden Dome opened in March 1988. During the opening ceremony, community leaders made televised predictions that the practice would lead to a reduction in crime.

- Did their efforts pay off? Interestingly, analysis of data from 1978 to 1991 showed that in March 1988, when the group size first surpassed this threshold of 100, the crime rate in Merseyside dropped significantly—by 13.4%. This reduction persisted until 1992, even as the national crime rate increased by 45%. Notably, Merseyside's crime ranking improved from the third highest among major metropolitan areas in 1987 to the lowest by 1992, with a rate 40% lower than predicted based on past trends. The study estimated that 255,000 fewer crimes occurred in Merseyside from 1988 to 1992 compared to what would be expected if it had followed national trends, yielding potential savings of over £1.25 billion in crime losses[27]. This significant reduction in crime rate cannot be explained by other factors, such as demographic shifts, economic changes, or police practices during the time period studied.

- In another study, researchers examined the impact of a large group practicing Transcendental Meditation and the TM-Sidhi program on violent crime in Washington, D.C. during June and July 1993. Washington, D.C. had suffered a 77% increase in violent crime between 1986 and 1992, despite the annual crime budget increasing to nearly $1 billion. Additionally, juvenile violent crime was projected to double in the following decade. To examine the impact of TM on reducing social stress and crime rates, about 4,000 participants gathered to increase collective coherence (harmony) by practicing TM. The research, monitored by an independent 27-member review board, analyzed weekly crime data from police records while accounting for factors like weather and crime trends. Results showed that homicides, rapes, and assaults (HRA crimes) dropped significantly—by up to 23.3%—during the project. This reduction was not observed during the same period in the previous five years. The analysis confirmed that the crime reduction could not be explained by increased police presence and that the effect persisted after the project ended[28]. Researchers estimated that a permanent group of 4,000 participants could reduce HRA crimes by 48% in the long run.

- These two studies represent just a fraction of the 600+ research studies conducted over the past five decades. These studies, carried out at over 200 universities and research institutions in 30 countries, have investigated the effects of TM on improving both individual and societal well-being, and they have consistently found similar positive outcomes[29]. How does TM improve the safety of cities? Scientists have found that Transcendental Meditation is linked with frontal-dominant alpha1 (8-10 Hz) electroencephalogram (EEG) activity, which reflects the brain's electrical activity. This alpha1 frequency in the brain's frontal cortex during TM practice is associated with a state of inner awareness and mental calmness. Scientists propose that when practicing in groups, meditation practitioners' coherent brain activities resemble those of laser light, in which photons are all in phase with each other, have the same wavelength, and travel in the same direction. Because of this laser-like coherence, researchers used the laws of physics to calculate the number of practitioners needed to see an effect in reducing the crime rate. This is how they derived the square root model. In this model, the TM-Sidhi program is believed to emit a coherent, orderly effect, similar to physical fields, that promotes harmony throughout the city.

The Mirror Rule

- Maybe a surprising example of how our minds can impact our living environment comes from the book *The Mirror Rule*. Based on a real story and written by Yoshinori Noguchi, this book explores the idea that our external reality reflects our inner world. The story follows Eiko Akiyama, a 41-year-old housewife struggling with her fifth-grade son, Yuta, who has been emotionally distant and troubled. Eiko is deeply frustrated, especially by the bullying her son is experiencing at school. In her despair, Eiko even has a fight with her husband, leaving her feeling lost and overwhelmed. She turns to her friend Yaguchi, who is knowledgeable about psychology, for advice. To her surprise, Yaguchi suggests that Eiko's strained relationship with her son might be connected to unresolved feelings toward her own father. He encourages her to list both her grievances and the things she's grateful for regarding her father and then to express her gratitude to him directly. Though initially skeptical, Eiko follows Yaguchi's advice. She writes a list of the positive and negative aspects of her relationship with her father. In doing so, she realizes that her father truly cared for her, despite her past focus on his flaws. She feels remorse and decides to confront her lingering pain and bitterness by forgiving him. In an emotional phone conversation, she thanks her father, releasing years of misunderstanding and resentment. After this act of forgiveness, Eiko's life undergoes a surprising transformation. Her son, previously withdrawn, opens up to her, sharing his school life and telling her that the bullying has stopped. Her husband also expresses gratitude toward her, and harmony is restored within her family. The book illustrates that our external circumstances often mirror our internal state. Harboring resentment can negatively affect our lives, leading to a reality filled with complaints and negativity. To change our reality, we must first transform our hearts.

Entangled and Connected: Embracing the Oneness

- How can we understand the impact of our minds on the environment? Quantum physicists may have found an answer: entanglement. Quantum entanglement is a fascinating and counterintuitive phenomenon that Einstein famously described as "spooky action at a distance." It refers to a state where two particles become so interconnected that a change in one instantly affects the other, regardless of the distance between them—even if they are billions of light-years apart. Einstein was skeptical about the existence of entanglement because nothing, including information, can travel faster than light. However, scientists have recently discovered that entanglement is not only real but also likely a fundamental mechanism of our universe.

- A fascinating paper published in *Nature* explores how quantum entanglement may underpin the very existence of our universe. The relationship between geometry and entanglement in quantum physics offers a way to understand how space itself might be "woven" together by this phenomenon. Imagine a 3D universe, which physicists call the "bulk," enclosed by a 2D "boundary," much like the skin of a balloon. The idea is that the activity within the 2D boundary—where quantum particles exist and interact—somehow generates the 3D universe we experience inside. Using mathematical models, physicists discovered that as they reduced the entanglement of particles on the boundary, the 3D space inside began to stretch and separate. If they removed all the entanglement, the space broke apart into disconnected pieces—similar to pulling apart a piece of chewing gum until it snaps[30]. This suggests that entanglement—the profound connection between particles, even when they're far apart—isn't just a strange quantum effect. It's potentially the "glue" that binds space-time together, making it smooth and continuous. Without entanglement, space itself could fall apart into isolated fragments. This finding indicates that entanglement is fundamental to the structure of the universe. It may even explain why our actions and thoughts can influence not only our bodies but also the environment we inhabit!

7.5 Mind Over Matter: The Science on the Power of Thought

Water Senses Your Thoughts

- In 2004, Dr. Masaru Emoto's book "The Hidden Messages in Water" became a New York Times bestseller. In this book, Dr. Emoto exposed water to various emotional stimuli, such as words, pictures, or music, and then examined the resulting ice crystals under a microscope. He found that the structure of the ice crystals was affected by the words and feelings directed toward the water. Water exposed to positive words and intentions, such as "thank you" and "I love you," formed beautiful, complex, and symmetrical crystalline structures. On the other hand, water exposed to negative words and intentions, like "I hate you" and "I want to kill you," formed disorganized and asymmetrical structures. Emoto also observed that different water sources produced different ice structures. For instance, water from a mountain stream produced geometric designs, while water from polluted sources produced distorted ice structures. Based on these observations, Emoto concluded that our thoughts and words can significantly impact water and the environment.

- Not surprisingly, the book garnered a lot of praise while also receiving significant criticism from many scientists for being biased and inconclusive. To test whether intention could influence the formation of water crystals, the Institute of Noetic

Sciences (IONS) in California, USA conducted a double-blind experiment. In this type of experiment, neither the person preparing the experiment nor the person analyzing the results knows which samples are the treated ones and which are the untreated controls until the experiment is complete. Researchers purchased four plastic bottles of water and randomly designated two as "treated" and two as "control." The treated bottles were placed in an electromagnetically shielded room at IONS in California, while the control bottles were stored separately. During a conference in Tokyo, Dr. Emoto led a group of 2,000 people in a five-minute prayer directed toward the treated water. The following day, scientists at IONS shipped all the bottles to Dr. Emoto's laboratory in Tokyo, where 50 water droplets from each bottle were analyzed for crystal formation. Scientists found that the treated water did indeed produce a slightly higher number of aesthetically pleasing crystals than the control group, and the treated crystals were judged to be significantly more appealing than those from the control group[31]. The scientists concluded that intention does appear to have a measurable impact on water crystal formation.

Influencing Random Events with Human Intention

- The most rigorous experiments exploring the mind's effect on physical reality were likely conducted by The Princeton Engineering Anomalies Research (PEAR) program. Established in 1979 by the Dean of Engineering, Robert G. Jahn, PEAR was a part of Princeton University's School of Engineering and Applied Science. The program spent nearly three decades investigating how human consciousness might interact with physical devices, systems, and processes. Since its inception, PEAR has published over 50 articles discussing both experimental results and the philosophical implications for science, engineering, and culture[32]. The program also developed theoretical models to better understand the role of consciousness in shaping physical reality.

- One major focus of the PEAR program was to examine human-machine interactions in an engineering context. Using a microelectronic random event generator (REG) that produced a randomly alternating sequence of positive and negative electrical pulses, researchers examined whether the intentions of human operators could affect the distribution of positive or negative events. Over 12 years, the study involved more than 100 human operators, over 1,000 experimental series, and more than 2 million trials using four different types of random devices. Interestingly, researchers found significant correlations between the operators' intentions and the mean outputs of different random binary devices, indicating a small (about 10^{-4} bits per bit processed) but significant and consistent deviation from chance (with $p \approx 3.5 \times 10^{-13}$)[33].

- The experiments also highlighted three main findings. First, researchers found notable differences between male and female operators. Although a few female participants achieved the highest individual scores, the overall correlation between intention and outcome was weaker for females than for males. Second, the physical distance between the person and the machine didn't seem to matter. Whether the operator was close to the machine or thousands of miles away, the effects were similar. Third, in a set of "off-time" experiments, the operators focused on the machine's operation at times different from when the data was actually being collected—sometimes days before or after. Surprisingly, researchers found that the results of these "off-time" experiments were very similar to those gathered in real time, including gender differences and some shifts in the data[33]. The only time researchers failed to see an effect of operators' intentions was when experiments were done using hard-wired devices or algorithmic approaches, methods known to produce pseudorandom (seemingly random but actually predetermined) events. This suggests that true randomness was necessary to observe the effects.

7.6 Consciousness After Death: Science or Fantasy?

Near-Death Experience: Consciousness Persisting Beyond Death?

- Dr. Eben Alexander, a highly trained neurosurgeon with appointments at esteemed medical schools including Harvard, Duke, the University of Massachusetts, and the University of Virginia, had no prior religious inclinations and firmly believed that consciousness was solely a product of brain activity. However, in 2008, Dr. Alexander contracted bacterial meningitis from an *E. coli* infection. The infection severely inflamed his brain and spinal cord, leading to brain shutdown and life-threatening conditions. He was placed in a medically-induced coma for seven days, and his Glasgow Coma Scale score indicated a complete absence of measurable consciousness.

- However, Dr. Alexander made a remarkable recovery from the coma. He claimed to have experienced consciousness outside of his body during this time, encountering what he described as an "angelic being" and communicating with a divine presence. His experience began in a dark, "primitive" realm, which he characterized as having a "scorched-earth" intensity. From there, a spinning, bright white light lifted him into what he called the "Gateway Valley." This realm resembled an earthly paradise, filled with lush landscapes, vibrant flowers, and cascading waterfalls. He encountered angelic beings, heard powerful chants and hymns, and felt enveloped in a profound sense of love and tranquility. He also described a connection to a higher, divine presence he referred to as "The Core," which he perceived as the source of all existence. This being conveyed a sense of infinite love and healing. The journey concluded when a boy told him he needed to return to his earthly life[34]. Dr. Alexander's near-death experience profoundly transformed him, leading him to believe that consciousness is not solely a product of the brain but transcends physical existence. He documented his experience in the book *Proof of Heaven*, published in 2012, which became a New York Times bestseller.

- Dr. Alexander's story triggered an intense debate. While it was well-received by individuals who have had similar near-death experiences, scientists and colleagues remained highly skeptical. Critics argue that his brain was not completely inactive, and his experience was simply a product of the erratic activity of his infected brain. However, Dr. Sam Parnia's research indicated that at least some aspects of the near-death experience are difficult to explain if brain activity is required for conscious experience.

- Dr. Sam Parnia, a critical care physician and researcher at New York University Langone School of Medicine, specializes in resuscitation science. He and his colleagues investigated what happens to consciousness during and after clinical death, specifically focusing on patients who experience cardiac arrest. Conducted across 15 medical centers in the U.S. and Europe, the study involved over 2,000 patients. Interestingly, Parnia's team found that up to 40% of revived individuals reported some level of awareness during cardiac arrest—a period when the heart stops, breathing ceases, and the brain shuts down, signaling clinical death. Some reported profound, life-changing experiences. Notably, 2% demonstrated full awareness and could accurately describe the events that occurred around them during their cardiac arrest[35]. These findings suggest that consciousness may persist even after clinical death. The implications are significant, challenging current views on death and the nature of consciousness and suggesting that it may not be exclusively tied to brain activity.

Reincarnation: Do We Live More Than Once?

- Another line of evidence for consciousness after death might come from reincarnation research. Dr. Brian Weiss, the author of the book "Many Lives, Many Masters", described how his patient 'Catherine' was able to recall her 86 previous lives during hypnosis therapy; some of the details (like the dates and specifics of certain historical events) were verified. However, critics argue that the past-life memories could be fabricated stories that happened to match real events by chance. In fact, researchers even conducted experiments and found that when children were told to make up their past-life stories, some details coincidentally matched real-world events. So, does this disprove reincarnation research?

- Dr. Ian Stevenson's findings may offer insights into this question. A former professor of psychiatry at the University of Virginia, Dr. Stevenson documented over 3,000 cases of children who reported memories of past lives. In addition to verifying the details of these past-life events through in-person interviews with related individuals, Dr. Stevenson focused on a specific characteristic of these children: birthmarks and birth defects. For instance, one case involved an Indian boy born with a deformed right hand, who claimed that in his previous life, he had undergone an amputation after injuring his right hand in a cutting machine accident. Another case involved an Indian man who remembered dying from a gunshot wound to the chest in a past life; he had painful birthmarks on his chest that resembled gunshot wounds. These statements were independently corroborated by medical records and the testimony of the attending physician, suggesting a potential link between birthmarks/birth defects and past-life traumas[36].

- While convincing, one major criticism Dr. Stevenson received was that his cases primarily came from countries like India and Sri Lanka, where belief in the afterlife is widespread. This raised concerns about potential bias and the possibility of fabricated stories. In response to this criticism, Dr. Stevenson expanded his research to include Western countries, including the United States. His student, Dr. Jim Tucker, continued this work after Dr. Stevenson's passing in 2007. In his book "Life Before Life," Dr. Tucker presented a compelling case from the US.

- John McConnell, the grandfather of a boy named William, was fatally shot during a robbery at an electronics store in 1992 in New York City. While trying to intervene, John was shot six times by a robber hiding behind the counter. Despite his injuries, John fought back before succumbing. William, born five years later, had pulmonary valve atresia, a heart defect that restricted blood flow to his lungs. He also had an underdeveloped right ventricle. These birth defects strikingly resembled the wounds his grandfather sustained. Furthermore, William displayed detailed memories of his grandfather's life, recalling John's nickname for a cat, "Boss," and specific events only John would have known. William also spoke about dying and being reborn, even correcting his mother about the dates of John's death and his own birth. His personality and habits mirrored those of his grandfather, suggesting a deep connection between them[37].

The Immortal Mind: A Theory

- If all these cases are genuine, how could consciousness persist after the brain's death? Dr. Tucker proposes that quantum mechanics might offer a theoretical explanation. According to quantum theory, matter exists only when observed. If consciousness is what brings the material world into existence, it might not depend on matter, including the brain. If this is the case, consciousness could potentially transition to a new brain and continue into another life after the death of the current body. This concept aligns with the views of physicists like Max Planck, who believed that the material world is a product of conscious observation.

- Additionally, the Swiss psychiatrist and psychotherapist Carl Jung, drawing inspiration from the ancient text "The Bardo Thodol" (also known as "The Tibetan Book of the Dead"), proposed that the human psyche contains a "collective unconscious"—a shared reservoir of archetypal symbols and experiences inherited from our ancestors. While consciousness might cease with the death of the physical brain, the collective unconscious, in this view, could persist beyond the constraints of space and time, carrying information from past lives that could be transferred to new individuals[38]. If this is accurate, the thoughts and behaviors of one's current life could influence the thoughts and behaviors of subsequent lives.

- Although many scientists find the studies on consciousness and past lives counterintuitive and difficult to accept, research into these mysteries continues to advance. Previously, scientists often avoided such topics, fearing potential damage to their reputations and careers. Fortunately, scientists today are increasingly open to exploring these areas, and now, more than ever, a comprehensive approach to research that integrates neuroscience, psychology, and quantum physics is essential. The increasing focus on the mind and past-life research holds promise for unlocking answers to some of the most profound questions about the universe and our existence within it.

Notes

1. Kam-Hansen, S., Jakubowski, M., Kelley, J. M., Kirsch, I., Hoaglin, D. C., Kaptchuk, T. J., & Burstein, R. (2014). Altered placebo and drug labeling changes the outcome of episodic migraine attacks. *Science Translational Medicine, 6*(218). https://doi.org/10.1126/scitranslmed.3006175

2. Tinnermann, A., Geuter, S., Sprenger, C., Finsterbusch, J., & Büchel, C. (2017). Interactions between brain and spinal cord mediate value effects in nocebo hyperalgesia. *Science, 358*(6359), 105–108. https://doi.org/10.1126/science.aan1221

3. Langer, E. J. (2009). *Counterclockwise: Mindful health and the power of possibility.* http://ci.nii.ac.jp/ncid/BB12235341

4. Park, C., Pagnini, F., Reece, A., Phillips, D., & Langer, E. (2016). Blood sugar level follows perceived time rather than actual time in people with type 2 diabetes. *Proceedings of the National Academy of Sciences, 113*(29), 8168–8170. https://doi.org/10.1073/pnas.1603444113

5. Aungle, P., & Langer, E. (2023). Physical healing as a function of perceived time. *Scientific Reports, 13*(1). https://doi.org/10.1038/s41598-023-50009-3

6. Crum, A. J., & Langer, E. J. (2007). Mind-Set matters. *Psychological Science, 18*(2), 165–171. https://doi.org/10.1111/j.1467-9280.2007.01867.x

7. Zahrt, O. H., Evans, K., Murnane, E., Santoro, E., Baiocchi, M., Landay, J., Delp, S., & Crum, A. (2022). Effects of wearable fitness trackers and activity adequacy mindsets on affect, behavior, and health: longitudinal randomized controlled trial. *Journal of Medical Internet Research, 25*, e40529. https://doi.org/10.2196/40529

8. Ranganathan, V. K., Siemionow, V., Liu, J. Z., Sahgal, V., & Yue, G. H. (2004). From mental power to muscle power—gaining strength by using the mind. *Neuropsychologia, 42*(7), 944–956. https://doi.org/10.1016/j.neuropsychologia.2003.11.018

9. Mendoza, D., & Wichman, H. (1978). "Inner" Darts: Effects of mental practice on performance of dart throwing. *Perceptual and Motor Skills, 47*(3_suppl),

1195–1199. https://doi.org/10.2466/pms.1978.47.3f.1195

10. Driskell, J. E., Copper, C., & Moran, A. (1994). Does mental practice enhance performance? *Journal of Applied Psychology*, *79*(4), 481–492. https://doi.org/10.1037/0021-9010.79.4.481

11. Toth, A. J., McNeill, E., Hayes, K., Moran, A. P., & Campbell, M. (2020). Does mental practice still enhance performance? A 24 Year follow-up and meta-analytic replication and extension. *Psychology of Sport and Exercise*, *48*, 101672. https://doi.org/10.1016/j.psychsport.2020.101672

12. *Books We Love: Meditation Saved My Life | Meditation Wise*. (n.d.). https://www.meditationwise.com/books-we-love-meditation-saved-my-life

13. McGreevey, S. (2019, September 12). Eight weeks to a better brain. *Harvard Gazette*. https://news.harvard.edu/gazette/story/2011/01/eight-weeks-to-a-better-brain/

14. Soon, C. S., Brass, M., Heinze, H., & Haynes, J. (2008). Unconscious determinants of free decisions in the human brain. *Nature Neuroscience*, *11*(5), 543–545. https://doi.org/10.1038/nn.2112

15. Creswell, J. D., Bursley, J. K., & Satpute, A. B. (2013). Neural reactivation links unconscious thought to decision-making performance. *Social Cognitive and Affective Neuroscience*, *8*(8), 863–869. https://doi.org/10.1093/scan/nst004

16. Sinclair-Gieben, A., & Chalmers, D. (1959). EVALUATION OF TREATMENT OF WARTS BY HYPNOSIS. *The Lancet*, *274*(7101), 480–482. https://doi.org/10.1016/s0140-6736(59)90605-1

17. Surman, O. S. (1973). Hypnosis in the treatment of warts. *Archives of General Psychiatry*, *28*(3), 439. https://doi.org/10.1001/archpsyc.1973.01750330111018

18. Shenefelt, P. D. (2000). Hypnosis in Dermatology. *Archives of Dermatology*, *136*(3). https://doi.org/10.1001/archderm.136.3.393

19. Ewin, D. M. (1992). Hypnotherapy for Warts (Verruca Vulgaris): 41 Consecutive Cases with 33 Cures. *American Journal of Clinical Hypnosis*, *35*(1), 1–10. https://doi.org/10.1080/00029157.1992.10402977

20. Weiss, B. (1988). *Many lives, many masters: The true story of a prominent psychiatrist, his young patient, and the past-life therapy that changed both their lives.* Touchstone USA.

21. Henry, R. C. (2005). The mental Universe. *Nature, 436*(7047), 29. https://doi.org/10.1038/436029a

22. Merali, Z. (2015). Quantum physics: What is really real? *Nature, 521*(7552), 278–280. https://doi.org/10.1038/521278a

23. Konvalinka, I., Xygalatas, D., Bulbulia, J., Schjødt, U., Jegindø, E., Wallot, S., Van Orden, G., & Roepstorff, A. (2011). Synchronized arousal between performers and related spectators in a fire-walking ritual. *Proceedings of the National Academy of Sciences, 108*(20), 8514–8519. https://doi.org/10.1073/pnas.1016955108

24. Feldman, R., Magori-Cohen, R., Galili, G., Singer, M., & Louzoun, Y. (2011). Mother and infant coordinate heart rhythms through episodes of interaction synchrony. *Infant Behavior and Development, 34*(4), 569–577. https://doi.org/10.1016/j.infbeh.2011.06.008

25. *Lovers' hearts beat in sync, UC Davis study says.* (2016, January 24). UC Davis. https://www.ucdavis.edu/news/lovers-hearts-beat-sync-uc-davis-study-says

26. University of Colorado at Boulder. (2017, June 21). When lovers touch, their breathing, heartbeat syncs, pain wanes, study shows. *ScienceDaily*. www.sciencedaily.com/releases/2017/06/170621125313.htm

27. Hatchard, G. D., Deans, A. J., Cavanaugh, K. L., & Orme-Johnson, D. W. (1996). The maharishi effect: A model for social improvement. Time series analysis of a phase transition to reduced crime in merseyside metropolitan area. *Psychology Crime and Law, 2*(3), 165–174. https://doi.org/10.1080/10683169608409775

28. Hagelin, J.S., Rainforth, M.V., Cavanaugh, K.L.C. et al. (1999). Effects of Group Practice of the Transcendental Meditation Program on Preventing Violent Crime in Washington, D.C.: Results of the National Demonstration Project, June-July 1993. *Social Indicators Research 47*, 153–201. https://doi.org/10.1023/A:1006978911496

29. Orme-Johnson, D. W., Cavanaugh, K. L., Dillbeck, M. C., & Goodman, R. S. (2022). Field-Effects of Consciousness: A Seventeen-Year study of the effects of

group practice of the Transcendental Meditation and TM-Sidhi programs on reducing national stress in the United States. *World Journal of Social Science, 9*(2), 1. https://doi.org/10.5430/wjss.v9n2p1

30. Cowen, R. (2015). The quantum source of space-time. *Nature, 527*(7578), 290–293. https://doi.org/10.1038/527290a

31. *Science of Water – Office Masaru Emoto.* (n.d.). https://masaru-emoto.net/en/science-of-messages-from-water/

32. *Princeton Engineering Anomalies Research.* (n.d.). https://pearlab.icrl.org/index.html

33. Jahn, R. G., Jr., Dunne, B. J., Nelson, R. D., Dobyns, Y. H., Bradish, G. J., Princeton Engineering Anomalies Research (PEAR), & School of Engineering and Applied Science, Princeton University. (1997). Correlations of Random Binary Sequences with Pre-Stated Operator Intention: A Review of a 12-Year Program. *Journal of Scientific Exploration, 11–11*(3), 345–367. https://pearlab.icrl.org/pdfs/1997-correlations-random-binary-sequences-12-year-review.pdf

34. *To do no harm | Magazine | The Harvard Crimson.* (n.d.). https://www.thecrimson.com/article/2020/2/20/eben-alexander-profile/

35. CBS News. (2017, October 20). *Researchers say there's evidence that consciousness continues after clinical death* [Video]. YouTube. https://www.youtube.com/watch?v=WnoIf2NwaRY

36. Stevenson, I. (1997). *Reincarnation and Biology: A contribution to the etiology of birthmarks and birth defects.* Praeger

37. Tucker, J. B. (2005). *Life before life: A scientific investigation of children's memories of previous lives.* St Martin's Press.

38. The SAP (Society of Analytical Psychology). (2023, November 30). Carl Jung | Biography, Archetypes & Beliefs | The SAP. *The Society of Analytical Psychology.* https://www.thesap.org.uk/articles-on-jungian-psychology-2/carl-gustav-jung/

Chapter 8
Concluding Remarks

This marks the end of the book. Throughout these pages, we've explored the wonders and mysteries of nature, the universe, the brain, and the mind. Each discovery is supported by scientific evidence, and I hope you've enjoyed the journey and found yourself as amazed as I have.

When I was a child, I was captivated by the mysteries of nature and the human mind. I couldn't help but ask "why": Why do the stars twinkle in the night sky? Why does the sun rise in the east? Why am I *me* and not someone else? For years, I pursued these questions, studying neural diseases and neuronal regeneration at a world-renowned institute in Boston under the mentorship of a Nobel Laureate. I hoped that one day, I would lead my own research lab, searching for the answers to these profound mysteries. Yet, I discovered that mainstream science—especially in the field of bioscience—often dismisses the role of the human mind in shaping physical reality. In an era where career success hinges on the number of publications and impact factors, the real questions, the ones worth dedicating a lifetime to, are often ignored or forgotten. Securing a job, winning grants, and attracting talented students have become more about metrics than about curiosity. That's why I decided to write this book. Not because we already have all the answers, but because it's time to refocus on the fundamental questions: *Who are we? What is the mind?* These are the mysteries that deserve our attention, and it's my hope that this book will inspire a renewed curiosity and an openness to exploring the unknown.

I didn't initially believe in spirituality. As a well-trained neuroscientist, I once firmly believed that consciousness was merely a byproduct of the material brain, and that all the secrets of life were encoded in our genes. I thought that understanding the function of each gene and how they interact would be enough to unlock the mysteries of life, including the workings of the mind. But one incident changed my perspective.

In 2011, my mother was diagnosed with rheumatoid arthritis (RA), an autoimmune condition where the body's immune system becomes overactive and starts attacking its own tissues, particularly the joints. RA is incredibly painful. At one point, her finger joints became so swollen that she couldn't manage daily activities like cleaning and cooking. For several months, a neighbor had to come over to help care for her. However, when I visited her six months later, she seemed fully recovered, and her finger joints looked no different from mine. I was shocked because I knew that there is no cure for RA—the best-known treatment is to suppress the overactive immune system and reduce inflammation with immunosuppressants. Curious, I asked her what "magic" medicine she had taken. To my surprise, she simply replied, "meditation." Then she looked at me and said, "If you want to improve people's lives, you need to learn more about the mind."

That moment was a turning point in my life. My mom is a Buddhist, and I remember how she would pray to Buddha whenever we faced difficulties when I was very young. After her divorce from my father, she worked tirelessly to earn enough money to raise my twin brother and me. With a monthly salary of less than $100 when I was a child, we frequently faced financial crises. But we managed to get through the hard times, perhaps due to her relentless prayers. Unlike my mom, I didn't have a particular attachment to any specific religion back then. I wasn't opposed to it either; I saw religion as a way to bring comfort during life's low points. But after witnessing her magic recovery from RA, I found myself, for the first time, genuinely curious about the power of the human mind. Was the effect real? How much could our mind influence our body? What was the mechanism behind this? If meditation could heal an "incurable" disease like RA, what other conditions might the power of the mind be capable of treating? Without hesitation, I began a journey of extensive research and reading. This book is the culmination of years of exploration and study, and, in my view, it has been far more fulfilling than completing my doctoral dissertation.

The journey has been rewarding but also full of challenges. The difficulties did not come from the learning itself, but rather from some friends and colleagues who believe I've taken a misguided scientific path, labeling the research findings I shared as pseudoscience. Yet, when we delved into the details of the experiments, they often found their criticisms losing ground and had to acknowledge that something intriguing was happening. As conventionally trained scientists, we are constantly under pressure to secure funding and publish research papers, which makes us accustomed to staying within our comfort zones and focusing on a single, narrow question at a time. I was always told by my mentor to be "specific" in my research questions. While this approach has undoubtedly contributed to the development of modern medicine and the current body of knowledge, it is unlikely to be effective in curing complex diseases. This is because, aside from a few cases where a disease is caused by a single gene mutation, the development and progression of complex diseases (such as Alzheimer's) involve hundreds, if not thousands, of genes. When we continue searching for individual genes responsible for the full spectrum of disease symptoms—a reductionist approach—we lose sight of the bigger picture. The numerous failures in Alzheimer's drug development, with only two out of 100 (2%) phase II and III clinical trials since 2003 showing any improvement, highlight the need to reflect on our methodology[1]. Finding the "switch" that governs the behavior of all these genes is crucial, and I believe that key may lie in our mind. My mom's recovery from RA and Phakyab Rinpoche's miraculous healing of his leg, avoiding amputation, are compelling examples of this possibility.

At the same time, the support from some of my fellow young scientists has been encouraging. They are genuinely inspired by research on the brain and mind, and they believe that mind-based therapy will be a critical part of future medicine. These young scientists share certain qualities: they are passionate about the unknown, open-minded, possess broad interests, and are deeply invested in the fundamental questions of nature and the universe. They represent the hope for our future. By combining the rigor of traditional science with an open-minded approach to unusual phenomena, we have a real opportunity to advance our understanding and develop a profound insight and respect for the universe we inhabit.

In fact, more and more medical practitioners are recognizing the importance of the mind in disease treatment and recovery. Mindfulness meditation has become a standard therapy in major hospitals like Massachusetts General Hospital, Stanford Health Care, and NewYork-Presbyterian. Leading companies like Google and Meta also dedicate resources to encourage meditation practices among their employees. Yet, we can't stop here, content with the idea that meditation simply makes us happier. We need to delve deeper into understanding the intimate connection between the mind, body, and universe. In an increasingly materialistic society, with the world facing rising levels of crime, wars, famines, and natural disasters, we must ask ourselves what we can do to address these problems. What legacy will we leave for the next generation? The answer, once again, might lie within the power of the human mind.

Finally, we must not forget the ancient wisdom of our ancestors. Whether it comes from the East or the West, this wisdom has withstood the test of time, shining through thousands of years of human history. I remember when I first began practicing meditation, my master told me, "You need to open your heart and care for other sentient beings more than yourself. Self-attachment is the root of all suffering, and helping other beings is the source of all happiness." When I asked why, he explained: "Because everything is a projection of your mind, and everything around you is interconnected. Loving and helping others is, in essence, loving and helping yourself." At first, I struggled to understand his words. But now, after years of learning, I see that scientists are slowly reaching similar conclusions. Is this a coincidence? Perhaps, or perhaps not. At first glance, science and religion may seem very different, even contradictory, but ultimately, they are different paths leading to the same truth. There may be countless ways to seek understanding, but the underlying truth remains constant. As Max Planck once said, "All matter originates and exists only by virtue of a force which brings the particle of an atom to vibration and holds this most minute solar system of the atom together. We must assume behind this force the existence of a conscious and intelligent mind. This mind is the matrix of all matter."

Let's purify our minds and make all the beautiful things happen.

P.S. Below, I've included a few powerful mantras my master taught me. These mantras can help to calm the mind, dispel negativity, improve health, and even attract prosperity. They should be used with good intentions—to better enable us to help others—to maximize their power. My master taught that praying with a narrow focus is acceptable, though in most cases, you're unlikely to gain more than what your past-life karma allows. I hope you find these teachings as beneficial as I have.

Mani Mantra: *Om Mani Padme Hung*

This mantra is for overcoming obstacles, bad fortune, illnesses, and disasters. Ultimately, it guides one toward liberation from suffering and the achievement of enlightenment.

Tara Mantra: *Om Tare Tuttare Ture Soha*

This mantra is for removing obstacles, fostering career development, and improving relationships with family, colleagues, and friends. Like the Mani Mantra, it also leads to liberation from suffering and enlightenment.

Manjushri Mantra: *Om Ah Ra Pa Tsa Na Dhi*

This mantra is for enhancing knowledge, mastering the highest levels of teachings, improving expressiveness and memory, and attaining ultimate wisdom.

May all beings benefit.

Note

1. Kim, C. K., Lee, Y. R., Ong, L., Gold, M., Kalali, A., & Sarkar, J. (2022). Alzheimer's Disease: Key Insights from Two Decades of Clinical Trial Failures. *Journal of Alzheimer's Disease*, *87*(1), 83–100. https://doi.org/10.3233/jad-215699

Stay Connected

Stay Curious and Informed!

Thank you for exploring the wonders within this book. Want to continue the adventure? Subscribe to my free bi-weekly newsletter for the latest breakthroughs and fascinating discoveries in science and beyond. Simply scan the QR code to join and never miss any inspiring innovations!

Scan to subscribe

Your Feedback Matters!

If you enjoyed this book, I'd be incredibly grateful if you'd take a moment to leave a positive review. Your thoughts inspire me and help others discover this work. Scan the QR code or use the link to share your experience. Thank you for your support—it means the world!

https://www.amazon.com/review/create-review?&asin=B0DPKL5J17

Made in United States
Cleveland, OH
12 November 2026

48209027R00078